Heavy Metals in Wastewater and Sludge Treatment Processes

Volume I
Sources, Analysis, and Legislation

Editor

John N. Lester, Ph.D.
Reader in Public Health Engineering
Department of Civil Engineering
Imperial College of Science and Technology
London, England

CRC Press, Inc.
Boca Raton, Florida

TD 758.5 .H43 H43 1987
v. 1

Heavy metals in wastewater
and sludge treatment
224126

Library of Congress Cataloging-in-Publication Data

Heavy metals in wastewater and sludge treatment
  processes.

  Includes bibliographies and indexes.
  Contents: v. 1. Sources, analysis, and legislation --
v. 2. Treatment and disposal.
  1. Sewage--Purification--Activated sludge process.
I. Lester, J. N. (John Norman), 1949-
TD758.5.H43H43   1987   628.3'57   87-10329
ISBN 0-8493-4667-3 (set)
ISBN 0-8493-4668-1 (v.1)
ISBN 0-8493-4669-X (v.2)

This book represents information obtained from authentic and highly regarded sources. Reprinted material is quoted with permission, and sources are indicated. A wide variety of references are listed. Every reasonable effort has been made to give reliable data and information, but the author and the publisher cannot assume responsibility for the validity of all materials or for the consequences of their use.

All rights reserved. This book, or any parts thereof, may not be reproduced in any form without written consent from the publisher.

Direct all inquiries to CRC Press, Inc., 2000 Corporate Blvd., N.W., Boca Raton, Florida, 33431.

© 1987 by CRC Press, Inc.

International Standard Book Number 0-8493-4667-3 (Set)
International Standard Book Number 0-8493-4668-1 (Volume I)
International Standard Book Number 0-8493-4669-X (Volume II)

Library of Congress Card Number 87-10329
Printed in the United States

FOREWORD FOR VOLUME I

The group of the transition and post-transition elements frequently referred to as "heavy metals" and generally accepted to include Cd, Cr, Cu, Hg, Ni, Pb, and Zn, plus the metalloids As and Se, have long been contaminants of sewages. With hindsight, it might have been preferable to have called the group "toxic elements"; this terminology permits the inclusion of elements such as aluminum which do not conform precisely to the physical or chemical meaning of "heavy metal" but which are nevertheless of concern for the same reasons. Even more appropriate might have been the description "exotic elements", a classification which could include Ag, Bi, Co, Mn, Mo, Sb, Sn, Te, Tl, and V. All of these elements can now be found in sewages, and in the environment generally, as a result of Man's activities. Although trade effluent controls have reduced the quantities of several of the toxic elements discharged to sewers, the increasing diversity of elements to be found in such discharges is testimony to the demands of technological innovation for new materials which previously had found little application.

That these elements are of concern in the aquatic environment is implicit in the plethora of legislation which has arisen in recent years. Most of the elements referred to here are classified as Priority Pollutants by the United States Environmental Protection Agency (EPA) and have been accorded the status of Dangerous Substances by the Commission of the European Communities (CEC). As such, limits have been placed on their concentrations in potable water, in sources of abstraction, and in effluent discharges. The behavior of the elements of concern in sewage treatment processes is of central importance in achieving compliance with the legislation, because their removal minimizes the pollution of surface waters. However, this is at the expense of their accumulation in the sludges produced. The disposal of these sludges is also controlled by legislation on a national and international basis.

Legislation is only of value if compliance can be checked; comprehensive legislation therefore demands a correspondingly comprehensive analytical capability. Recommended standard methods in Europe and the U.S. invariably specify atomic absorption spectrophotometry as the current method of choice, but little more than a decade ago, relatively insensitive, imprecise colorimetric methods were still in use. Now, a variety of methods are available, but many are not suited to wastewater matrices. Despite the recognition by Kirchoff and Bunsen of absorption in flames in the mid-19th century, it was not until a century later that the analytical possibilities of atomic absorption were described by Walsh in his 1953 patent application. By 1966, the technique using flame atomization was well-established. Although L'vov had designed a carbon tube furnace capable of greater sensitivity in 1961, it was the simpler design of Massman in 1968 that led to the appearance of several commercial designs a year later. The much greater sensitivity permitted not only the determination of a wider range of elements as low concentrations, but provided the necessary detection limits for metal speciation.

It is remarkable that perhaps the most difficult matrix from an analytical point of view, sewage sludge, was the subject of some early attempts at metal speciation. This was certainly not because sewage sludge was considered an easy matrix to work with, but becasue there was a perceived need to obtain data amenable to biological interpretation so that the potential adverse effects of sludge use in agriculture could be quantified. The most expedient way of obtaining such data, that of chemical extraction, has acknowledged limitations and has not found widespread routine application. However, extraction techniques have contributed to an understanding of metal speciation in sewage sludges, the importance of which is now reflected in sludge disposal guidelines which make reference to factors such as cation exchange capacity, pH, and soil composition which can affect the bioavailability of metals.

Considerable interest is currently being shown in metal speciation in natural waters.

Sewages and sewage effluents have received much less attention, probably because the analytical problems are more challenging in these matrices. However, one aspect of metal speciation in biological sewage treatment, that of metal binding to bacterial extracellular polymers, has stimulated interest in the mechanisms of metal immobilization generally and how this is affected by metal speciation. There is no doubt that there is considerable scope for the development of analytical speciation methods for sewages and related matrices. This will form the basis of a broader understanding of the behavior and significance of heavy metals in wastewaters and in the wider context of the hydrological cycle.

THE EDITOR

John N. Lester, Ph.D. is Reader in Public Health Engineering, Civil Engineering Department, Imperial College of Science and Technology, London.

Dr. Lester graduated from the University of Bradford, with a B.Sc. degree in applied biology in 1971; he was awarded the M.Sc. of the University of London and the Diploma of Imperial College (DIC) in 1972 for studies in biochemistry and plant physiology. In 1975 he received his Ph.D. from the University of London for research in the field of environmental microbiology. After a Postdoctoral-Fellowship in the Public Health Engineering Laboratory of the Civil Engineering Department at Imperial College, he joined the academic staff as a lecturer in 1977. In 1986 he was appointed Reader in Public Health Engineering.

Dr. Lester is a member of the Institution of Water Engineers and Scientists, Institution of Public Health Engineers, Institute of Water Pollution Control, and the Institute of Biology. He has been a member of the Technical Programme Committee of several International Conferences, including the Second (London 1979), Third (Amsterdam 1981), and Fourth (Heidelberg 1983) Heavy Metals in the Environment Conferences.

He has published more than 200 research papers, presented numerous lectures at international and national conferences, and written or contributed to several books.

While Dr. Lester is continuing research into the behavior of metals in wastewaters, sewage sludges, and surface waters, he also continues a significant research program on the behavior of organic micropollutants in similar systems. In addition, he is actively involved with the anaerobic treatment of high strength industrial wastewaters and transport-related air pollution.

CONTRIBUTORS

Peter W. W. Kirk, Ph.D.
Postdoctoral Fellow
Department of Civil Engineering
Imperial College of Science and
 Technology
London, England

Donna Lee Lake, Ph.D.
Postdoctoral Research Assistant
Department of Civil Engineering
Imperial College of Science and
 Technology
London, England

Thomasine Rudd, Ph.D.
Environmental Scientist
Consultants in Environmental Sciences,
 Ltd.
London, England

Tom Stephenson, Ph.D.
Senior Lecturer in Biochemical
 Engineering
North East Biotechnology Centre
Department of Chemical Engineering
Teesside Polytechnic
Middlesbrough, England

Robert Sterritt, Ph.D.
Lecturer
Department of Civil Engineering
Imperial College of Science and
 Technology
London, England

TABLE OF CONTENTS

Volume I: Sources, Analysis, and Legislation

Chapter 1
Scope of the Problem 1
T. Rudd

Chapter 2
Sources of Heavy Metals in Wastewater 31
T. Stephenson

Chapter 3
Pollution Control Legislation 65
P. W. W. Kirk

Chapter 4
Determination of Heavy Metals in Wastewater Matrices 105
R. M. Sterritt

Chapter 5
Chemical Speciation of Heavy Metals in Sewage Sludge and Related Matrices 125
D. L. Lake

Chapter 6
Physical and Electrochemical Speciation 155
R. M. Sterritt

Index 175

TABLE OF CONTENTS

Volume II: Treatment and Disposal

Chapter 1
Primary Mechanical Treatment 1
J. N. Lester

Chapter 2
Biological Treatment 15
J. N. Lester

Chapter 3
Sludge Treatment.. 41
T. Rudd

Chapter 4
Water Treatment and Reuse 69
T. Stephenson

Chapter 5
Sludge Disposal to Land 91
D. L. Lake

Chapter 6
Sludge Disposal to Sea.. 131
P. W. W. Kirk

Index 145

Chapter 1

SCOPE OF THE PROBLEM

T. Rudd

TABLE OF CONTENTS

I.	Definition and Classification of Heavy Metals	2
II.	Chemical Properties and Speciation	4
III.	Toxicity and Human Health Effects	9
	A. Arsenic	11
	B. Cadmium	12
	C. Lead	13
	D. Mercury	14
	E. Nickel	15
	F. Chromium	16
	G. Selenium	17
	H. Zinc	18
	I. Copper	18
	J. Aluminum	19
	K. Beryllium	19
	L. Thallium	19
	M. Silver	20
	N. Tellurium	20
	O. Vanadium	20
IV.	Reasons for Concern	21
References		23

I. DEFINITION AND CLASSIFICATION OF HEAVY METALS

There is considerable dissension as to the actual definition of a heavy metal and consequently over which elements should be regarded as such. This is demonstrated by the variety of different criteria listed by Neiboer and Richardson.[1] One of the more common definitions is that heavy metals constitute those with specific gravities of approximately 5 g/cm^3 or greater.[2] The group includes metals from the transition series and Groups IIA, IIIB, IVB, VB, and VIB of the periodic table. However, the term heavy metals is frequently applied where there are connotations of toxicity, and thus a less vigorous definition of the group, normally used in the environmental context, is that it includes lighter metals such as aluminum and beryllium and metalloids such as arsenic, selenium, and antimony. Metalloids behave chemically both as a metal and a nonmetal; thus arsenic, although not a metal in the strict chemical sense, is linked toxicologically to the heavy metals because of the sulfur-sequestering properties common to both.[3] Jarvis[4] included certain metalloids in his definition of heavy metals as those "generally considered to be of sufficient environmental distribution and/or abundance as to be in some way environmentally or biologically significant as a toxic substance".

Various classifications of heavy metals have been made, based on either their chemical or biological properties. Chemical classifications have largely arisen from the traditional concepts of acids and bases, in which the electron-rich donor or Lewis base provides electrons for the electron-poor acceptor or Lewis acid to form a covalent bond. Pearson[5] categorized metals (electron-accepting acids) into hard and soft acids or acceptors (Table 1), depending on the strength of their binding with respective electron donors and hence the magnitude of their equilibrium constants for complex formation. Hard acceptors were observed to associate with hard donors and soft acceptors with soft donors (Table 1). Examples of such associations given by Förstner and Wittman[6] are the natural occurrence of the hard acceptor calcium as chalk ($CaCO_3$) and that of the soft acceptor mercury as the ore cinnabar (HgS), where carbonate and sulfide are hard and soft donors, respectively.

Biological classifications have been based on the essentiality and/or toxicity of individual metals. Essential elements form either macro- or micronutrients, depending on the relative concentrations in which they are physiologically required. The majority of the heavy metals fall into the latter category, occurring at 1 to 2 mg/kg tissue.[7] Table 1 contains groups of essential and nonessential metals, although it should be noted that several heavy metals previously thought to be nonessential, such as nickel and vanadium, have fairly recently been reclassified, largely due to improved analytical instrumentation, allowing determination of progressively lower metal concentrations in living tissues. Distinctions such as those in Table 1 should thus be regarded more as state-of-the-art evaluations than absolute categorizations.

Caution should also be exercised when defining a metal as toxic since effects such as toxicity are largely a function of concentration, thus certain elements such as arsenic and selenium may be beneficial at low doses but become hazardous when in excess of physiological requirements[8] (Figure 1). The difference in concentration may be low, as for boron in plants and selenium in animals;[10] the latter may be deficient at dietary intakes of 0.04 mg/kg or less, while toxicity becomes evident at intakes of 3 to 4 mg/kg or higher, a difference of only two orders of magnitude.[11] Moreover, when present in excess, some of the essential elements such as selenium and vanadium can exert greater toxicity than nonessential elements such as mercury and thallium.[12]

A classification proposed by Wood,[13] however, divided the gross effects of the potentially hazardous elements of the periodic table into categories of "noncritical", "very toxic and relatively accessible", and "toxic but very insoluble". The relevant elements in each of these groups are listed in Table 1 for comparison with the previous classifications. The

Table 1
CLASSIFICATION OF SELECTED METALS AND ASSOCIATED LIGANDS

Classification				Ref.
Hard acceptor/acid	**Intermediate**	**Soft acceptor/acid**		
K^+, Na^+, Be^{2+}, Ca^{2+}, Mg^{2+}, Mn^{2+}, Al^{3+}, As^{3+}, Co^{3+}, Cr^{3+}, Fe^{3+}	Co^{2+}, Cu^{2+}, Fe^{2+}, Ni^{2+}, Pb^{2+}, Zn^{2+}	Ag^+, Au^+, Cu^+, Tl^+, Cd^{2+}, Hg^{2+}		5, 14
Hard donor/base	**Intermediate**	**Soft donor/base**		
H_2O, OH^-, F^-, Cl^-, SO_4^{2-}, CO_3^{2-}, O^{2-}, PO_4^{3-}	Br^-, NO_2^-, SO_3^{2-}	SH^-, RS^-, CN^-, SCN^-, S^{2-}, CO, R_2, S, RSH		5, 14
Class A metals	**Borderline**	**Class B metals**		
K^+, Na^+, Ba^{2+}, Be^{2+}, Ca^{2+}, Mg^{2+}, Al^{3+}	Cd^{2+}, Co^{2+}, Cr^{2+}, Cu^{2+}, Fe^{2+}, Mn^{2+}, Ni^{2+}, Pb^{2+}, Sn^{2+}, V^{2+}, Zn^{2+}, Fe^{3+}, $As(III)$, $Sb(III)$, $Sn(IV)$	Ag^+, Cu^+, Tl^+, Hg^{2+}, Bi^{3+}, Tl^{3+}, $Pb(IV)$		1
Class A ligands	**Borderline**	**Class B ligands**		
H_2O, OH^-, NO_3^-, F^-, SO_4^{2-}, CO_3^{2-}, O^{2-}, HPO_4^{2-}, PO_4^-, ROH, $RCOO^-$, ROR	Br^-, Cl^-, N_3^-, NO_2^-, O_2^-, SO_3^{2-}, O_2^{2-}, O_2, NH_3, N_2, RNH_2, R_2NH, R_3N, $CONR$	H^-, I^-, R^-, CN^-, RS^-, S^{2-}, CO, R_2S, R_3As		
Essential	**Possibly beneficial**	**No apparent metabolic function**		
Animals: Co, Cr, Cu, Fe, Mn, Mo, Ni, Se, Sn, V, Zn Plants: B, Cu, Fe, Mn, Mo, Se, Zn	As, Ba	Bi, Cd, Hg, Pb, Tl		10, 15
Noncritical	**Toxic/accessible**	**Toxic/insoluble**		13
Al, Ca, Fe, K, Mg, Na	Ag, As, Be, Bi, Cd, Co, Cu, Hg, Ni, Pb, Sb, Se, Sn, Te, Tl, Zn	Ba		13

Note: Some elements in the original classifications have been omitted for clarity.

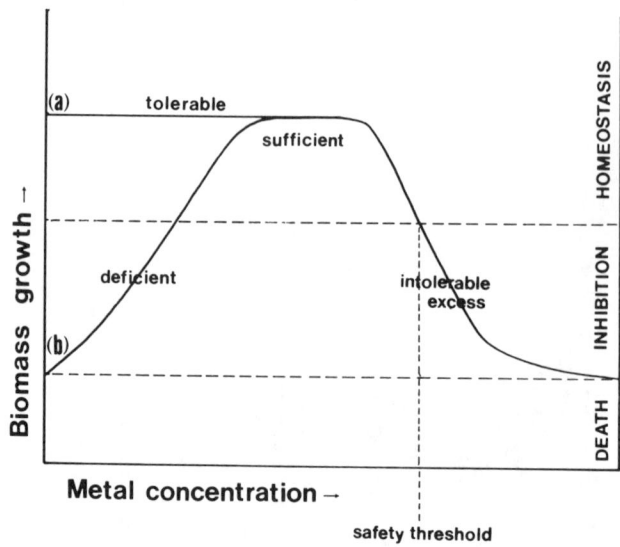

FIGURE 1. Effects of increasing metal concentration on biomass for (a) a nonessential element such as cadmium and (b) an essential element such as copper.[9]

apparent correspondence of the first category with that of Pearson's[5] hard acceptors and similarities between the "very toxic and accessible" metals and the soft acceptors was observed by Förstner and Wittman.[6] These authors postulated a tentative connection between the chemical and biological classifications by pointing out that the toxic soft acceptors were likely to bind to soft donors such as the sulfydryl groups in proteins. Since sulfydryl groups form active sites on proteins, their blockage through heavy metal binding could give rise to toxic effects.

Such interrelationships between the chemical and biological behavior of heavy metals have since been formulated to give a further classification designed to facilitate the interpretation of metal ion toxicity on a biochemical basis. Nieboer and Richardson[1] have proposed that metals should be designated Class A, B, or borderline, depending on their relative affinity for O-, S-, and N-containing ligands (Table 1). Class A metals, generally corresponding to the macronutrients, have an electron donor preference for oxygen, while Class B metals prefer nitrogenous or sulphurous groups. The borderline metals, comprising the first row of the transition elements and generally corresponding to the class of micronutrients, form stable complexes with all types of ligand. In terms of toxicity, metals thus classified decrease in the order: Class B, borderline, Class A. The degree of toxicity exerted by each group is dependent on its relative ability to bind to incorrect sites, or to block or displace other requisite ions from active sites on biological molecules.

II. CHEMICAL PROPERTIES AND SPECIATION

In addition to the relative affinity of the free metal ion for various biological functional groups, it has become increasingly recognized that toxicity greatly depends on the form or species of metal to which an organism is exposed.[16,17] In the hydrological system, metal speciation is determined by a combination of the chemical properties of the individual metal and the moderating effects of the external environmental conditions which are imposed upon it. Possible species which may occur are illustrated in Table 2.

As an example, nickel has a $3d^8 4s^2$, or a $3d^{10}$ electron configuration, neither of which

Table 2
AQUATIC METAL SPECIES[12,18]

Physicochemical form	Example	Phase	Approximate diameter (nm)
Free metal ion	$Cuaq^{2+}$	Dissolved	0.8
Inorganic complexes	$CuCO_3$, $CuOH^+$, $CU(CO_3)_2$, $Cu(OH)_2$	Dissolved	1—2
Organic complexes	(Cu chelate structure with $CH_2-C=O$, NH_2, O groups)	Dissolved	2—4
Colloids and large polymers	Organic/inorganic		
Surface bound, adsorbed	>Fe–OCu; $-C(=O)-O-Cu$	Particulate	10—500
Solid, bulk-phase lattice	CuO, $Cu_2(OH)_2CO_3$	Particulate	>450

is stable. To achieve stability, two electrons are removed to attain the $3d^8$ configuration, giving a predominant oxidation state of Ni(II). In natural aerobic waters at neutral pH, nickel is octahedrally coordinated as $(Ni[H_2O]_6)^{2+}$ and ionic nickel is likely to constitute 90% of the total metal.[19] Nickel also forms complexes with ligands such as $OH^+ > SO_4^{2-} > Cl^- > NH_3$, while its solubility under anaerobic conditions is determined by the presence or absence of sulfide.[20] Nickel speciation in surface waters is affected by the presence or absence of suspended solids, although estimates of the proportion of total nickel associated with particulates vary[21,22] from 7 to 99%. Snodgrass[23] has calculated that 0.5% of nickel in the major world rivers is in solution, 3.1% is adsorbed, 47% exists as a precipitated coating, 14.9% is associated with organic material, and 34.4% is present in crystalline forms. In wastewaters, nickel is usually present as Ni(II) and forms stable soluble complexes, such as with cyanide discharged from electroplating industries.[24] Although it has a lower affinity for biological wastewater materials than other metals such as copper, nickel has been shown to form complexes preferentially with soluble organic ligands in wastewater effluents[25] and this may contribute to its generally high mobility in aqueous systems.

Like nickel, the metals copper, cadmium, and zinc also form stable octahedral hydrated ions in oxygenated water, but differ in other respects. Copper occurs in aqueous solution in oxidation states of (I) and (II), but the former is unstable above equilibrium concentrations of 10^{-2} M and in aerated natural waters is generally oxidized to copper(II). The aqueous chemistry of copper(II) is dominated by its great tendency to form complexes, as indicated by the magnitude of its stability constants for complex formation in comparison with other metals (see Sillen and Martell[26]). The proportion of free ionic to total dissolved copper in natural waters is approximately 1% and considerably less in waters with a high organic load and above neutral pH, such as sewage effluents.[27] In the absence of organic ligands, hydrolysis and precipitation control copper speciation. Predominant forms are the free ion below pH6 and carbonate[28] between pH6 and 9.3, while at pH values typical of natural waters, copper hydroxide and basic carbonate form colloidal suspensions and precipitates.[29]

Cadmium has properties intermediate between those of mercury and zinc. It occurs in an oxidation state of (II) and the ionic form normally predominates in unpolluted soft waters of relatively low pH value.[30] Cadmium can form a range of soluble complexes such as with carbonate, sulfate, chloride, and hydroxide ligands and the synthetic chelating agents EDTA and NTA. The least-soluble cadmium salt in aerated natural water is cadmium carbonate, which precipitates between pH 8.5 and 11.[31] However, humic complexes are significant in organically loaded waters such as sewage effluent, while the greatest removal of cadmium from solution is likely to be via adsorption onto solids, particularly organic particulates. This propensity explains its relatively efficient removal during sewage treatment.[32,33]

The normal valence states of zinc are metallic (O) and zinc(II). The octahedral stereochemistry of the hydrated ion $(Zn[H_2O]_6)^{2+}$ deviates in the presence of ligands to produce complexes with coordination numbers commonly of four and five.[34] Zinc(II) has amphoteric characteristics and forms complexes with anionic, cationic, neutral, or more complex ligands. Organozinc compounds such as ZnR_2 are also formed, where R = methyl, ethyl, n-propyl, n-butyl, n-pentyl, or vinyl groups, for instance. The stability of zinc complexes in natural water tends to be lower than that of metals such as copper, cadmium, and lead.[28] Practical difficulties in resolving the complexity of the aqueous behavior of zinc, in combination with its ubiquity as a contaminant, have produced some dissension as to the types of zinc species predominating in surface waters, particularly seawater.[16] However, in natural aerobic water, a significant fraction of zinc is likely to consist of simple nonionic forms; thus, in addition to the hydrated ion, inorganic compounds (e.g., $ZnCO_3$), stable organic compounds (e.g., Zn-cysteinate), and forms adsorbed or occluded onto organic (Zn^{2+}-humic acid) or inorganic (Zn^{2+}-clay) colloids exist.[35] At pH values of approximately 6 in freshwater, the noncolloidal inorganic forms predominate, but at pH > 7, adsorption to both organic and inorganic materials increases significantly.[34]

Lead can exist in several valence states; Pb(O) as metal, Pb(I), Pb(II), and Pb(IV), and occurs naturally as the sulfide, carbonate, sulfate, and chlorophosphate. The most common oxidation state is Pb(II), soluble salts of which dissolve to give the hydrated Pb^{2+} ion. Solubility calculations indicate that lead solubility is likely to be less than 1 $\mu g/\ell$ in a system containing[36] lead, water, carbon dioxide, and sulfur at pH 8.5 to 11. Above pH6, lead solubility is controlled by pH and dissolved carbon dioxide; at constant pH, it decreases in inverse proportion to alkalinity, while below pH 6.5, it is likely to increase.[36] The majority of lead salts are sparingly soluble, but since lead in natural aerated water is normally undersaturated in terms of precipitation of insoluble salts, the major mechanism controlling its solubility in both fresh and saline waters is adsorption onto suspended particulates.[37] Lead in sewage effluents has been shown to be associated with suspended solids[38] and may also exist as soluble or colloidal complexes formed with organic ligands such as the sulfydryl, carboxyl, and amine functional groups present in living matter.[39]

The solubilities of some inorganic complexes of heavy metals are shown in Table 3. These data indicate that the nature of the metal complexes formed with different ligands can influence the solubility of a compound and hence the potential bioavailability and toxicity of the constituent metal. This may be of considerable significance, as in the case of nickel where there is an inverse relationship between metal salt solubility and carcinogenicity, with the least soluble salt, nickel subsulfide, being the most carcinogenically active.[40] The valency or oxidation state of the metal also has considerable influence on its bioavailability. Table

Table 3
SOLUBILITY OF SOME HEAVY-METAL COMPOUNDS[41]

Anion	OH⁻	S²⁻	Cl⁻	CO₃²⁻	NO₃⁻	SO₄²⁻
Cd^{2+}	SS	SS	1400	SS	1500	760
Cr^{3+}	SS	Dec	SS		S	Dec
Co^{2+}	SS	SS	640	SS	VS	362
Cu^{2+}	SS	SS	730	SS	1220	205
Fe^{2+}	SS	SS	644a	0.06	S	S
Fe^{3+}	SS	Dec	Dec		S	Dec
Hg_2^{2+}		SS	0.002	SS	Dec	0.6
Hg^{2+}		SS	69	SS	VS	Dec
Mn^{2+}	SS	SS	723a	0.0065	VS	630
Ni^{2+}	SS	SS	642	0.09	VS	370
Zn^{2+}	SS	SS	4320a	0.01b	1170	540

Note: Solubilities are given as 10^3 times the mass of the anhydrous solute per mass of water at 293 K. VS, very soluble; S, soluble; SS, sparingly soluble; Dec, decomposes on addition to water.

a At 298 K.
b At 288 K.

3 indicates that iron(II) salts are largely soluble, whereas the majority of iron(III) salts are insoluble. Mercury occurs in three oxidation states: (O) as elemental mercury, (I) as mercurous compounds, and (II) as mercuric compounds. The mercurous salts are less soluble than the mercuric and are consequently less toxic.[39] Hazards can arise, however, from the discharge of the less harmful species of mercury, since aquatic sediments can readily oxidize metallic mercury to Hg^{2+}, and certain microorganisms such as *Clostridium cochlearium* can convert Hg^{2+} to mono- and dimethylmercury by biomethylation. Methylated species of mercury are even more toxic than the mercuric ion.[16] Other metals which can be biomethylated by bacteria or fungi, frequently those species found in wastewater environments, include lead,[42] cadmium and tin,[43] selenium and tellurium,[44] and arsenic.[45] Although the products of methylation may be more toxic than the free metal, they are often volatile and can be released to the atmosphere.

Arsenic compounds are ubiquitous and are mostly insoluble in water. Arsenic commonly exists in either the trivalent or pentavalent state, associated with both organic and inorganic ligands. Under most conditions in natural waters, inorganic forms of arsenic prevail, with the more thermodynamically stable arsenate predominating in aerated waters and arsenite occurring at low pH and under reducing conditions. The trivalent, inorganic arsenicals are more hazardous than the pentavalent forms. Chromium can exist in various valency states, but the most common in water are chromium(III) and chromium(VI). The hexavalent form exerts the greater toxicity; therefore its conversion to the trivalent form, which may be precipitated as insoluble compounds, is highly desirable.[46] This has been shown to occur during the process of wastewater treatment by Brown et al.[47]

Barium occurs naturally as one of two insoluble salts: barium carbonate or barium sulfate. Although its other salts are soluble in water and acid and can exert considerable toxicity, barium is frequently adsorbed or precipitated rapidly and thus removed from solution. An example of the inverse relationship between solubility and toxicity is barium sulfate, the insoluble nature of which renders its use acceptable for coating the human alimentary tract to increase the contrast on x-ray photographs.[46] Beryllium is unlikely to be found in hazardous concentrations in unpolluted natural waters. Its chloride and nitrate salts are very soluble in water, whereas the hydroxide and carbonate are insoluble. Selenium, which like arsenic exhibits both metallic and nonmetallic properties, occurs as basic iron(III) selenite, calcium selenate, or elemental selenium. Its terrestrial availability is localized in areas where seleniferous soils occur, and aquatic concentrations reflect this distribution. Elemental selenium is insoluble until oxidized to selenite or selenate.[39]

Thallium occurs in either a monovalent or trivalent state, the former predominating in aqueous media. Thallium(I) resembles the alkali earth metal cations in chemical behavior, its hydroxide salt is soluble while the sulfide, chloride, and chromate are sparingly soluble. Inorganic complexes of thallium(III) are more water soluble, but its organic complexes are more stable than those of thallium(I). Thallium(III) can also be methylated, but since inorganic and organic forms of this metal exert a similar degree of toxicity, this property is perhaps of less environmental significance than for mercury.[48]

Other properties of metals which affect their speciation and potential bioavailability have been identified by Jackson,[49] who demonstrated that the propensity of cadmium, copper, iron, mercury, and zinc to form complexes with organic carbon in lake sediment could be correlated with metal electronegativity ($r = 0.992$; $p = <0.001$), the equilibrium constant for metal ion hydrolysis ($r = 0.93$; $p = 0.01$ to 0.02), and ionic radius ($r = 0.966$; $p = 0.0001$ to 0.01). In contrast, Steinberg[50] found that the proportions of the alkali earth metals calcium and magnesium which were chelated with dissolved organic carbon in stream water had an inverse relationship with ionic radii of the metals. Jackson[49] further observed that heavy metals, such as mercury, which display strong tendencies to form stable covalent bonds with ligands, are preferentially associated with organic material and sulfide, although local environmental conditions dictate which of the properties of a metal dominate its speciation under specific conditions.

Snodgrass[23] has summarized those factors which affect the speciation of metals in aqueous systems. Mechanisms which influence the proportions of soluble metal species include:

1. Acid-base interactions
2. Oxidation-reduction reactions
3. Metal-inorganic complexation
4. Metal-organic complexation

Those that affect solid-phase speciation are

1. Direct adsorption onto inorganics, e.g., clays
2. Adsorption or coprecipitation onto hydrous iron and manganese oxides, which subsequently adsorb to clays
3. Adsorption or complexation with biological particles such as detritus or bacteria
4. Association with organic macromolecules, e.g., humic and fulvic acids, which subsequently adsorb to clays
5. Ion exchange with clay colloids
6. Precipitation

The rate and extent of the above speciation reactions are controlled by environmental parameters such as pH and Eh, ionic strength, the type and concentration of inorganic and organic ligands, and the presence of solid surfaces for adsorption. By altering such parameters, and consequently the speciation of metals in a wastestream, wastewater treatment can have a significant effect on metal bioavailability and potential toxicity.[51] The process of anaerobic digestion in particular has been shown to have a stabilizing effect on metal forms;[52,53] for example, via the precipitation of metal sulfides.

III. TOXICITY AND HUMAN HEALTH EFFECTS

The ultimate cause for concern about heavy metals in the environment is their extreme toxicity towards man. The toxicity of a metal or metal compound has been defined as its intrinsic capacity to cause injury, including potential carcinogenic, mutagenic, and teratogenic effects.[54] Metal toxicity may be manifested in either acute or chronic forms, with acute toxicity encompassing the total adverse effects produced by a toxicant administered[9] as a single dose or as multiple doses over a period of ≤24 hr. Acute or short-term metal toxicity is commonly evaluated in animals by using standard tests such as the LD_{50}, the lethal dose required to kill 50% of a population of test organisms in a fixed time period, e.g., 48 or 96 hr. A measurement of potential toxicity, similar to that of pH for hydrogen ion concentration, has been postulated by Luckey and Venugopal,[55] where pT is the negative log of the molar concentration of the toxic substance. Campbell et al.[56] have specified pT ranges for heavy metals of low (pT = <0.5), moderate (pT = 5.0 to 7.0), and high (pT = >7.0) inherent toxicity.

Although short-term or acute toxicity testing may be of use in assessing risks from single high exposures, of greater environmental concern with regard to heavy metals are the effects of long-term exposure to low concentrations of toxicant. This form, chronic toxicity, is difficult to assess under laboratory conditions due to the long experimental duration required. The relevance of experiments in which test organisms are exposed to a consistent concentration of toxicant over a long time period is also questionable, since concentrations of toxic metals in the environment tend to be subject to fluctuation.

Other aspects of chronic toxicity which may not be immediately apparent from laboratory-based studies on groups of test organisms include the way in which chronic toxicity may reduce an organism's general ability such that it becomes vulnerable to other potentially lethal factors such as infection. If not killed by chronic exposure to a metal, organisms may still suffer sublethal effects such as a reduction in reproductive success. An example of this was reported in a French community living near an open cast mine whose birthrate decreased by one child per family following establishment of the mine, in comparison to the birthrate of a nearby unpolluted community. This was tentatively attributed to elevated levels of lead and cadmium (caused by the mining operations) in the river used for irrigation, and subsequently in the community food crops.[57] Effects such as these, when evident in an entire population, may be more hazardous to the species as a whole than the loss of a number of individuals by direct intoxication.[58]

Toxicity testing on laboratory organisms can provide valuable information on the effects of metals in isolation under acute or chronic conditions, but the results of such tests cannot be extrapolated directly to man. Apparently safe intakes for humans can only be inferred. Species susceptibility to toxicants such as heavy metals differ considerably, as evidenced by the widely disparate tolerance of sheep and pigs to copper,[11] and differences in the distribution and metabolism of metals makes the extrapolation of animal data to humans very difficult.[59] Since toxicity testing obviously cannot be carried out on human subjects, the health effects of heavy metals and other hazardous chemicals have largely been assessed through epidemiology. Epidemiology is the study of the distribution and determinants of disease and injury in human populations[60] and involves the collation of evidence from incidences of population exposure to elevated environmental concentrations of specific metals from natural or industrial sources. Two of the most well-documented epidemiological studies into the effects of overexposure to heavy metals were those involving cadmium and mercury poisoning in Japan, first reported in 1955 and 1956, respectively.

Itai-itai disease, caused by cadmium poisoning, was initially diagnosed in farm workers in northern Japan, who habitually drank water from the Jinzu River and consumed rice grown in paddy fields irrigated by the same river. The Kamioka mine, situated on the upper reaches of the river, discharged cadmium into the river via a faulty wastewater-treatment system. The cadmium contained in the ore mined for lead and zinc production was disposed of since it had no utility value and consequently accumulated by sedimentation from the river water, which was diverted from flowing in the main channel to standing static on the rice fields. Over periods of 30 years, the farm workers accumulated high concentrations of cadmium in their bodies, up to 15,000 mg/kg in bones in one extreme case.[61] The disease was named Itai-itai ("ouch ouch") due to the pain caused by the decalcification and final fracturing of bones symptomatic of cadmium poisoning. Up until 1965, it was estimated that 200 people, mostly farm women over middle age, had suffered from the disease and, in more than half of these cases, it had proved fatal.

Minimata Disease was first observed in Minimata Bay in southwest Japan, and by 1959 had been attributed to consumption of fish contaminated with mercury originating from effluent from a Chrisso Corp. chemical factory. The factory released relatively nontoxic inorganic mercury together with toxic methyl mercury as a result of using mercuric sulfate catalysts for acetaldehyde production. Further transformation of the inorganic species of mercury to organic forms occurred in the sediments of Minimata Bay following discharge of the effluent. The methyl mercury was accumulated by fish and shellfish in the bay and subsequently in the local Japanese people for whom fish formed a significant proportion of the diet. In the entire area contaminated by methyl mercury, there have been 784 officially designated Minimata disease cases, of which 103 have died, 21 within a year of the onset of the disease.[62] Many people are still suffering from subacute and chronic symptoms of the disease, such as numbness of limbs, hand tremors, and sensory disturbances.

Epidemiological surveys are of value because they examine the effects of toxicants *in situ* rather than in isolation in the laboratory, thus any exacerbating or ameliorating influences exerted by the environment are taken into account. However, there are several factors inherent in human epidemiological studies which limit their usefulness:[54]

1. If exposure has been ubiquitous, it may be impossible to assess the effects of a metal because there is no unexposed control group.
2. It is often difficult to determine doses in cases of human exposure.
3. It is usually hard to identify small variations in common effects within an entire population, such as behavioral alterations.
4. Interactions within an exposed population existing under natural rather than laboratory conditions cannot be controlled.

The determination of threshold levels for heavy metals, that is, the minimum level below which toxic and carcinogenic effects do not occur, is difficult and has caused some controversy.[63] This has been evident particularly with regard to carcinogens; current opinion is that there are no threshold limits for this category of toxicant.[64] Underwood[11] has listed the factors which are capable of potentiating the toxicity of heavy metals. These include the stage of life cycle, age, species, and physiological state of the exposed organism, the mode of entry of the toxicant into the exposed organism, the physicochemical form or species of the metal to which the organism is exposed, and the extent to which other elements or compounds interact environmentally and metabolically with the toxicant. Metabolic interactions may be synergistic (where the total effect of two or more metals or compounds is greater than the sum of their individual effects), additive (where the total effect is equal to the sum of the individual effects), or antagonistic (where the presence of one metal or compound reduces the toxic effect of another). Factors such as these mean that no safe maximum tolerance of a metal can be designated to cover all possible circumstances.

The preliminary mode of action by which heavy metals exert toxicity in man and animals is by reacting with sulfur-donor atoms of proteins, resulting most commonly in enzyme deactivation. In addition, by replacing other essential elements like calcium and magnesium, heavy metals can destabilize the structure of biomolecules. In the case of nucleic acids, such a combination of reactions may lead to faulty replication and result in genotoxic and mutagenic effects producing heritable genetic disorders and cancers. The chemical nature of some metals and metal species further permits transplacental movement inducing embryotoxicity and teratogenicity. The following sections assess the major toxic effects of some of the more significant heavy metals.

A. Arsenic

Toxic effects caused by inorganic and aliphatic species of arsenic differ considerably from those caused by organic arsenic compounds, which makes reliable indexes of human exposure difficult to establish. Tri- and pentavalent inorganic forms predominate in water, but the majority of in vivo reactions are attributed to the trivalent form.[65]

Soluble forms of arsenic such as sodium arsenite are readily absorbed by the body, whereas arsenic trioxide and less soluble arsenicals are poorly absorbed and largely excreted in the feces. Organic forms, which may be formed in vivo, are virtually completely absorbed following ingestion. Absorbed arsenic affects tissues rich in oxidative systems, such as the alimentary tract, liver, kidney, lung, and epidermis. Skin concentrations increase following exposure, whereas liver and kidney levels decrease. Individuals consuming drinking water containing 0.41 mg/ℓ of arsenic were found to have higher concentrations in the blood, hair, and nails than an unexposed population.[66] Ingestion of arsenic is likely to cause subacute and chronic dermatological effects such as skin hyperpigmentation, keratosis, and cancer. Absorption of inorganic arsenic from the gastrointestinal (GI) tract is rapid, causing diarrhea and vomiting within a few minutes in cases of acute poisoning.[67] Other clinical signs manifested in acute cases include fever, emaciation, anorexia, depression, and hair loss. The effects of arsenic can be reduced by selenium, which acts antagonistically, probably by competitive inhibition at active sites for arsenic.[68]

Such active sites for trivalent arsenic, the more toxic form to humans, include sulfydryl ($-$SH) groups of proteins; its binding inhibits the activity of sulfydryl enzymes and thus blocks fat and carbohydrate metabolism and cellular respiration.[65] Pentavalent arsenic uncouples oxidative phosphorylation and competes with phosphorus in metabolic reactions.[46] Arsenic carcinogenicity testing in animals has proved negative, but epidemiological studies implicate arsenicals, particularly trivalent species, in human cancer either as carcinogens or cocarcinogens.[69] Becking[70] tentatively attributed the differing carcinogenetic response between man and animals to the fact that man is the only species which excretes monomethylarsonic acid as a major metabolite. However, it has been suggested that arsenic is an essential element and may in fact reduce cancer induction.[71]

Arsenic has been implicated in genotoxicity, through its ability to induce inhibition of nucleic acid and protein syntheses; arsenite prevents DNA repair.[69] In addition, it has the potential to be fetotoxic in humans, although organic arsenicals do not cross the placenta, anionic forms do.[72] A premature baby, born to a woman who had ingested arsenic trioxide, was found to contain high concentrations of arsenic in its liver, kidneys, and brain and died shortly afterwards.[73]

Several case studies involving chronic arsenic poisoning due to the naturally occurring high levels of arsenic in drinking water have been detailed by Lafontaine.[67] In Argentina, well water containing from 0.9 to 3.4 mg/ℓ of arsenic resulted in symptoms of symmetrical palmar and planar hyperkeratosis appearing 2 to 3 years after the onset of exposure; while in Taiwan, where levels of 0.4 to 0.6 mg/ℓ of arsenic were recorded in drinking water, an increased prevalence of skin cancer, hyperpigmentation, and keratosis was observed. In northern Chile a quarter of a million people have been exposed to levels of arsenic in drinking water from 0.05 to 0.96 mg/ℓ (mean 0.6 mg/ℓ). Following the installation of a treatment system, this mean value dropped to 0.08 mg/ℓ with a simultaneous decrease in the incidence of chronic arsenical disease. The World Health Organization (WHO) has set drinking-water standards based on such epidemiological data,[74,75] although the U.S. Environmental Protection Agency (EPA) considers that no definite threshold concentration can be derived[76] and instead states levels "which may result in incremental increase of cancer risk over the lifetime".

B. Cadmium

Human health data relating to cadmium have been extensively reviewed and discussed.[77-80] The absorption of ingested cadmium is generally low in humans, ranging[80] from 4.7 to 7%, but may increase in cases where dietary calcium is low.[46] Absorption is species dependent, with cadmium sulfide and selenide being taken up to a lesser degree than chloride and oxide forms. Cadmium tends to concentrate in the wall of the small intestine — probably in the mucosal cells — but the routes of cadmium transport from the intestine are not well-documented.[82] Erythrocytes have been shown to contain cadmium, and approximately 60% of the metal is probably bound to metallothionein, a metal-binding protein of low molecular weight.[83] It is generally accepted that steady-state cadmium blood levels reflect exposure, but humans newly exposed to cadmium may require 100 days for equilibrium in the blood to be attained.[82]

Once adsorbed, cadmium is tenaciously retained in man and accumulates mainly in the liver and kidney, with the concentration being directly related to the oral dose administered.[84] There is no known physiological requirement for cadmium, and there appears to be no homeostatic mechanism by which the body can maintain cadmium at a constant safe level.[39] Cadmium concentration in the renal cortex increases with age until about 50, when the concentration levels off or decreases. The biological half-life of cadmium in the kidneys has been estimated at 10 to 40 years,[85] and in the liver, at 5 to 10 years. Excretion is mainly via the urine with a contribution from feces, saliva, and hair, but only a very small quantity is excreted daily.[82] In humans with average body burdens of 10 to 60 mg, cadmium is excreted in the urine at about 0.5 to 2.0 µg/ℓ.[77]

The main target organ following acute oral exposure to cadmium is the GI tract, while the kidney, bone, heart, pancreas, testes, and hematopoietic system are affected following chronic exposure. After prolonged exposure, the kidney becomes the critical organ. Excessive accumulation of cadmium in the renal cortex eventually leads to a decreased reabsorption of proteins in the proximal tubules, resulting in tubular proteinuria, an increased excretion of low-molecular-weight proteins in the urine.[77] Anemia is frequently observed in chronic cadmium toxicity, which results from the metabolic antagonism between cadmium and iron and cadmium and copper.[11] Normal metabolism is also disrupted through competition and displacement of zinc in metalloenzymes by cadmium.

Symptoms of exposure to low concentrations include vomiting, diarrhea, and colitis, while continuous exposure causes hypertension, heart enlargement, and death. Determination of cadmium in hair has been proposed as an indicator of hypertensives.[87] The renal dysfunction caused by cadmium reduces the production of active vitamin D which mediates calcium absorption and thus induces osteomalacia. Such skeletal thinning and bone breakage was observed in victims of the widespread cadmium poisoning incident in Japan described earlier.

Parenteral administration of cadmium to laboratory animals has induced cancer, but evidence of carcinogenic effects in man has been inconclusive. Limited mutagenic assays suggest that cadmium is nonmutagenic, although an interactive effect with a known mutagen has been demonstrated.[70] Inhibition of nucleic acid synthesis and protein synthesis can be induced by cadmium, among other metals.[69] In teratogenic studies on animals, zinc and selenium were found to be antagonistic towards cadmium, whereas lead and mercury were synergistic. No known teratogenic or embryotoxic effects have been observed in humans,[72] but cadmium can cross the placenta to be retained in the fetus. Cadmium levels in the umbilical cord were higher than those in the maternal blood of pregnant Japanese women.[88]

Attempts have been made to estimate the threshold exposure for cadmium. Based on preliminary autopsy studies, it has been concluded that the critical level of cadmium in the renal cortex for the appearance of proteinuria is about 200 mg/kg wet weight. A metabolic model applied to this data did not provide an exact dose-response relationship, but a daily intake in food and drink of 0.2 to 0.4 mg of cadmium would appear to be sufficient to attain the critical level in the kidney.[77] Threshold levels of 0.2 to 0.248 mg cadmium per day for an adult human have also been estimated for oral intake.[77] Concentrations in unpolluted water are generally low, less than 1 $\mu g/\ell$, and constitute only a small fraction of daily intake. No cases of toxicity arising from oral intake of cadmium in water have been reported in respect to nonindustrial exposure.[11]

C. Lead

The human toxicology and epidemiology of lead exposure have been reviewed by the World Health Organization,[89] the Department of Health and Social Security,[90] and the Royal Commission on Environmental Pollution.[91] The occurrence of lead in drinking water may be the result of contamination of the water source or corrosion in the distribution system where lead piping is used. Lead has no known physiological function. The absorption of ingested lead from the GI tract is less than that by inhalation and amounts to approximately 10% in adults and 53% in children, with 18% retained.[89] Absorption varies with age and is affected by differences in the chemical species of lead, the presence or absence of food, and the hardness of the water supply.[91] In Canada it has been estimated that drinking water in hard-water areas can contribute 15%, and in soft-water areas 46%, of the lead in the diet.[92] Once absorbed, lead passes into the bloodstream where more than 95% is bound to the erythrocytes,[93] causing increased fragility and reduced life span of the cells. Direct determination of lead in venous or capillary blood is the most widely used assessment of lead in the body.[91]

Lead has a high affinity for bone; in the steady state, about 90% of the adult body burden is bound to this tissue. The concentration of lead in bones increases throughout most of life. The average male body burden has been quoted as 165 mg, with occupationally exposed workers accumulating up to 566 mg.[94] Lead has a biological half-life in man of many years, but to quantify this period accurately is extremely difficult, largely due to the differing availabilities of lead stored in the bones and in soft tissue. Estimates vary from 19 to 27 days for blood lead and 10 to 20 years for lead in bone.[36] The relative importance of different sources and pathways of lead may vary considerably between groups of people. Excretion of lead from the body is largely via the urine (approximately 76%) and the GI tract (approximately 16%); the remaining 8% is excreted through sweat, hair loss, and nails.[89]

Acute or classical lead poisoning in human adults is manifested by anemia (caused by the metabolic antagonism of iron at the absorptive level), alimentary symptoms (such as constipation and colic), wrist and foot drop, renal damage, and sometimes encephalopathy. Symptoms in children include irritability, loss of appetite, occasional vomiting, intermittent abdominal pain, and constipation. If the poisoning is unchecked, vomiting becomes more persistent, muscle coordination is affected, and coma may result. It is extremely difficult to state a threshold below which acute effects do not occur, since considerable variation is evident. Levels of more than 0.8 mg/ℓ have been considered to be unequivocal evidence of clinical lead poisoning,[95] but occupational exposure screening occasionally reveals men with a blood lead concentration in excess of 1.5 mg/ℓ with no obvious clinical symptoms, while symptoms are sometimes encountered with a blood lead concentration of 0.6 mg/ℓ or less.[91]

There is considerable uncertainty regarding the effects of lead at low blood lead concentrations and as to whether or not a threshold exists below which damage does not occur, although concentrations of 0.25 to 0.4 mg/ℓ have been proposed in this respect.[95] Most chronic effects of lead are nonspecific in nature, which makes it difficult to establish cause-effect relationships. Lead has been associated with behavioral effects and intellectual impairment in children, but the interpretation of epidemiological evidence relating to these phenomena is made difficult by the number of variables and associated factors to be considered. A number of fundamental physiological processes are known to be affected by lead. For example, lead inhibits several stages of the synthesis of heme (a precursor of hemoglobin) by inhibiting the enzyme δ-aminolevulinic acid dehydratase (δ-ALAD). Measurement of δ-ALAD in urine and serum is used as an indication of the biological effects of lead. Oral administration of lead salts has induced cancer in laboratory animals but there is no evidence of carcinogenic effects on man.[91] Lead is embryotoxic and teratogenic in humans causing spontaneous abortions. An incident involving acid drinking water containing lead from pipes in Scotland resulted in miscarriage, fetal death, and abnormal births.[97]

The combination of chemically aggressive water supplies and lead piping have caused a number of cases of lead poisoning; in one area of France, 150 cases were recorded in 8 months in 1983. A survey of 321 residents of the Vosgian region in France found blood lead levels of 0.23 mg/ℓ in men and 0.15 mg/ℓ in women, and identified a threshold limit of 0.02 mg/ℓ in water below which consumers' blood lead was stable and above which it increased commensurately with lead concentrations in water.[98] An epidemiological study of 8500 people in the U.K. in 1977 found highest blood levels related to plumbosolvent water. Lower blood lead levels determined in 1981 confirmed the effectiveness of remedial water treatment.[99]

The Royal Commission on Environmental Pollution[91] considered that the exposure of the general population to lead should be reduced, particularly when it is considered that the mean blood lead concentration in the general population is only about one quarter of that at which symptoms of classical poisoning may occasionally occur. In the U.S., the National Research Council stated that the present U.S. drinking-water limit for lead "may not, in view of other sources of environmental exposure, provide a sufficient margin of safety, particularly for fetuses and young growing children", and it was suggested that the limit be lowered.[100] Two approaches to the formulation of drinking-water guidelines have been discussed by the World Health Organization.[101] One is to estimate the contribution made by drinking water to the level of lead in blood, and the second is to take account of the provisional tolerable weekly intake of 3 mg lead per adult established in 1972.[102]

D. Mercury

The toxicity of mercury varies considerably depending on the species to which an individual is exposed. The least toxic forms are inorganic mercurials, which are poorly absorbed from the GI tract, with between 5 and 15% of the ingested dose being taken up.[103] Phenylmercury exerts a similar degree of toxicity, since, following absorption, it is readily broken down to inorganic species in the tissues. Organic mercury, however, is absorbed to 80 to 100% of that ingested, although a high proportion is excreted in the bile. Once absorbed, inorganic mercury accumulates in the kidney and liver, with the renal cortex containing the highest concentrations. The biological half-life of a single oral dose of inorganic mercury has been reported to be between 30 and 60 days, excretion being mainly via the urine.[93] Organic mercury is accumulated in tissues. Its lipid solubility permits it to diffuse through cell membranes, resulting in appreciable concentrations occurring in the brain which may thus contain 10 to 20% of the total methylmercury body burden. The affinity of organic mercury for nervous tissue in particular accounts for many of its observed health effects. Its biological half-life in the body is 70 to 74 days, with elimination occurring mainly through the feces and urine.[103] Intake of inorganic mercury has been shown to cause release and redistribution of organic species within the body, implying that they may share similar binding sites, for which inorganic mercurials have higher affinity.[104] Methylmercury is a soft acceptor exhibiting a high capacity for binding to sulfydryl groups in proteins, and it has been suggested that relative stabilities of different methylmercuric complexes may have bearing on their individual toxic properties.[105] There is evidence that the toxicity of mercurials may be antagonized by selenium, as mine workers apparently unaffected by abnormally high mercury levels in their tissues were found to have coaccumulated selenium in an approximate 1:1 molar ratio.[106]

Acute oral exposure to inorganic mercury in water results in intense local corrosion of the GI tract and subsequent renal failure.[93] Chronic mercury poisoning causes neurological disorders with parasthesia of the extremities, tremors, headaches, irritability, vertigo, and depression. Where methylmercury has been ingested, damage to the cerebellum and sensory pathways results in progressive incoordination, loss of sight and hearing, and mental deterioration from toxic neuroencephalopathy.[11] The main target organ for organic mercury poisoning is the central nervous system (CNS). Observations have been made on the retardation of nerve conduction in an occupationally exposed worker.[107]

Mercury has not been shown to be a carcinogen, but of all the metals, it is probably the most potent teratogen. The ease of crossing the placental barrier is greatest for organic mercury, which exerts more toxicity following cumulative chronic exposures than a single acute dose. Accumulation in the fetus occurs by the "fetal trap", where mercury crosses the placenta by facilitated transfer and becomes firmly bound to receptors. This forms a concentration gradient, resulting in up to 28% higher mercury concentrations in the fetal blood than the maternal blood.[72]

Both adult and fetal casualties were reported in large numbers for the two major epidemics of mercury poisoning in humans. The long-term toxic effects of the mercury pollution of Minimata Bay and subsequent bioconcentration through the food chain to humans were a 38% mortality in the affected population and a higher ratio of newborn to pregnant victims. This indicated the susceptibility of the fetuses compared to the mothers, some of whom only had symptoms of numbness and fatigue, while their offspring suffered brain damage, spasticity, and blindness. Similarly, in the largest incidence of mercury poisoning, which occurred in Iraq in 1956 and 1960 due to consumption of grain treated with a mercury-based fungicide, there were 6530 victims and 459 hospital deaths. Thirty-one of the victims were pregnant, of whom fourteen died. Children born to these women had blood mercury levels of up to 2500 mg/ℓ and suffered severe brain damage.[72] Estimation of the health risks arising from mercury have been mainly focused on methyl-mercury compounds because of the environmental transformation of inorganic species through methylation, and the greater toxicity of the organic form. The World Health Organization[108] concluded, on the basis of epidemiological evidence, that the fetus is the stage of the human life cycle most sensitive to methylmercury. However, estimates of total mercury in blood and hair have only been made for the most sensitive group in the adult population. The earliest observable effects in 5% of this group would be expected at a blood mercury concentration of 200 to 500 μg/ℓ and a hair concentration of 50 to 125 μg/g, equivalent to a long-term daily intake of 3 to 7 μg/kg body weight.[108] Such estimates are based on the premise that parasthesia is the earliest detectable effect of methylmercury intoxication, although experiments with laboratory animals suggest that histological changes in the nervous system precede such severe effects.[109] The World Health Organization[102] Expert Committee on Food Additives established a provisional tolerable weekly intake of 0.3 mg of total mercury per person, of which no more than 0.2 mg should be present as methyl mercury (expressed as mercury); these amounts are equivalent to 5 and 3.3 μg, respectively, per kilogram of body weight.

E. Nickel

The toxicity of nickel and its compounds has been examined by Sunderman,[110] Nielsen,[111]

Mushak,[112] and Bencko.[113] While nickel has been shown to be essential for normal growth in some animals, its role as an essential element for man has not yet been confirmed.[114] Absorption of nickel from the GI tract is high compared to that of other metals, but constitutes probably 3 to 6% of nickel ingested in water.[115] Once absorbed, nickel is transported in the bloodstream bound to serum albumin; mean nickel serum values of unexposed humans are of the order of 2 to 3 $\mu g/\ell$.[116] It has been suggested that the high, species-specific solubility of different nickel compounds in human blood serum could be used as a method of assessing their relative toxicity.[117] During distribution, nickel becomes complexed with amino acids, peptides, and proteins and accumulates in tissues of the lung, kidney, liver, endocrine gland, and brain, depending on the mode of absorption. It is rapidly excreted from the body in urine (0.7 to 5.2 $\mu g/\ell$) and also via sweat (up to 52 $\mu g/\ell$) and hair (up to 220 $\mu g/kg$).

The most toxic form of nickel known to man is nickel carbonyl, and the majority of epidemiological data on nickel poisoning are based on occupational exposure to this form in, for example, nickel refineries.[118] When inhaled, nickel carbonyl can cause severe lung damage and produces symptoms of headaches, vertigo, nausea, vomiting, and insomnia.[119] Oral exposure, however, produces a low toxicological response due to the low absorption via this route, and appears to be tolerated at low levels via a homeostatic mechanism. Concentrations of nickel in tissues of exposed workers have been shown to decrease over several weeks once they had left the contact area.[120]

Nickel is capable of causing mutagenic effects by binding to phosphates and heterocylic bases of nucleic acids in the place of other elements such as magnesium. This results in destabilization of the double helical structure of DNA or RNA and unfaithful replication.[69] Although nickel inhalation has been documented as causing cancer of the nasal cavity and lung,[121] there is little evidence to suggest that nickel in drinking water may be carcinogenic. No gametotoxic or teratogenic effects or interference in human development have been observed in connection with nickel exposure.[112] Maximum admissable concentrations of nickel in drinking water have been specified by the Council of the European Communities (0.05 mg/ℓ), but not by WHO or EPA.[122]

F. Chromium

Chromium is an essential element since one species forms the glucose tolerance factor.[123] It is generally present in the environment as poorly soluble trivalent salts of low toxicity. Absorption of chromium(III) salts by the GI tract is low in humans, averaging from 0.5 to 1%; the presence of food generally retards absorption. Hexavalent chromium, the most toxic form, is more readily absorbed but is seldom present.[124] Mammals can tolerate up to 200 times the total body burden of approximately 6 mg without harmful effects. The concentration of chromium in the body decreases with age, but its half-life is not known with certainty. Urine is the main excretory route for chromium (7 to 10 $\mu g/day$) and would appear to represent the most reliable index of current exposure.[124]

Toxic effects of chromium are related to its chemical form. Most effects seem to be caused predominantly by chromium(VI) compounds, although this may be related to the limited uptake of chromium(III) compounds. Acute effects from oral exposure include severe corrosion of the GI tract and kidney necrosis, while chronic exposure via inhalation of chromium(VI) compounds can result in chrome ulcers in the skin and corrosive holes in the nasal septum. Localized effects include acute irritative determatitis[125] and increased allergic sensitivity.[114]

The hexavalent form of chromium is recognized as a mammalian carcinogen;[69] evidence from in vitro tests suggests that direct mutagenic initiation of somatic cells via DNA interactions is involved.[70] Recent studies with human red blood cells indicated that although chromium is taken up as the hexavalent form, it is reduced intracellularly to chromium(III), which in a complexed form ultimately becomes the genotoxic agent.[126]

A case study on Croatian workers exposed to chromium recorded incidences of skin sensitivity and ulcers, perforation of the nasal septum, and significantly higher occurrence of respiratory cancer than in a control group.[127] An incidence of large-scale chromium(VI) poisoning in Japan in 1975, cited by Förstner and Wittman,[6] involved more than 30 deaths due to contamination of drinking water. Chromium had leached from chemical wastes used as building materials into groundwater, and, as a result, locally abstracted drinking water was found to contain concentrations of more than 2000 times the official threshold limit.

Schroeder and Lee[128] suggested that the possible occurrence of oxidation of chromium(III) and reduction of chromium(VI) under natural conditions necessitated basing water-quality standards on total chromium rather than hexavalent chromium concentrations. In contrast, the U.S. National Academy of Sciences[129] recommended that "regulations governing the presence of chromium in drinking water distinguish between the nutritionally useful trivalent and the more toxic hexavalent form". However, such a proposition is complicated by a number of uncertainties, including analytical difficulties in determining the hexavalent form alone and the likely conversion of trivalent to hexavalent chromium in drinking-water sources under oxidizing conditions such as during chlorination.[64,101]

G. Selenium

Toxicological effects of selenium have been reviewed by the National Academy of Sciences[130] and specific effects have been discussed by Spallholz et al.[131] Selenium is an essential element, forming part of the enzyme glutathione peroxidase and other biological compounds.[132] However, the difference between essential and toxic levels is small in comparison to other essential metals such as copper and zinc. The valency of selenium affects its toxicity, which appears to be exerted through mechanisms such as replacement of thiol groups by SeH groups.[6]

The extent of absorption of selenium by man is uncertain, but selenium is present in human blood and in all samples of human urine analyzed by sensitive detection methods.[133] In an Oregon study, the blood of laboratory personnel and athletes was found to contain 150 to 300 $\mu g/\ell$ of selenium. The concentration in blood appeared to be independent of smoking habits, sex, exercise, and general fitness.[130] Frost[134] and Schroeder et al.[135] have estimated that the average human diet in the U.S. gives a selenium-consumption rate of 1.8 mg/month (i.e., 25 μg per kilogram body weight per month), which is comparable with a requirement for selenium of 20 μg per kilogram body weight per month for an adequate diet for rats.

Toxic effects of selenium ingestion have largely been observed in animals, but some human cases have been noted. Selenosis in man is manifested by chronic dermatitis, loss of hair, loss, discoloration, and brittleness of finger nails, and dental caries. However, the evidence presented for these effects does not establish that selenium was the cause.[130] Carter[136] reported a case of acute lethal selenium poisoning of a 3-year-old boy whose symptoms included dilation and increased permeability of the peripheral vasculature and congestion and edema of the lungs and GI tract. Severe acute toxic effects and chronic effects have been reported in livestock following consumption of selenium-accumulating plants. The effects observed appeared to increase gradually as selenium intake increased. The extrapolation of results from animals to man must, however, be made with caution.[129]

Conclusions regarding any possible carcinogenicity of selenium have largely been based on demographic studies and comparisons of levels of selenium in the blood of patients with and without malignancies. Several studies have suggested that human cancer mortality might bear an inverse relationship to selenium distribution,[137,138] and this has been supported by results from controlled animal studies. However, epidemiological evidence of this nature has inherent flaws and more research is required.[71] Selenium has been demonstrated to be teratogenic in animals, causing shortening of the limbs. Reputed effects on humans include four abortions and the birth of a child with clubbed feet in a group of occupationally exposed women, and a high incidence of neural tube birth defects in Ireland where soil selenium concentrations are high.[72]

Frost[134] compared maps of early heart mortality and cardiovascular related deaths for various areas of the U.S. and demonstrated an inverse relationship between ambient selenium levels and mortality pattern. Selenium also appears to be required for the optimal functioning of vascular endothelium in animals. It has been suggested that selenium may protect against toxicity of other heavy metals, such as cadmium, lead, mercury, and silver,[139] while arsenic is an antagonist of selenium. Under most circumstances, however, it seems that selenium does not protect laboratory animals by increasing the excretion of heavy metals but rather causes an accumulation of metal in tissues. Selenium has been shown to divert the binding of cadmium and mercury from low-molecular-weight proteins to high-molecular-weight ones. The selenide form of selenium could also react with metals to form insoluble metal selenides, thus reducing metal toxicity.[139]

The National Academy of Sciences concluded that an adequate and safe intake level of selenium was between 50 and 200 µg/day,[64] while the WHO concluded that a daily intake of 130 to 200 µg approximates normal intake figures published.[101] The latter stated that no more than 10% of the dietary intake should be derived from drinking water. Based on a consumption of 2 ℓ/day, selenium in drinking water should therefore not exceed 0.01 mg/ℓ.

H. Zinc

Zinc is an essential element, with a human average daily requirement of 10 to 20 mg. Most zinc is obtained from the diet with water usually supplying only a small fraction of the intake. Zinc is absorbed in the duodenum via a binding ligand; normally 20 to 30% is absorbed, although the proportion varies with zinc status, protein and calcium in the diet, and age of the organism. Zinc is required for metalloenzymes and has a role in synthesis and metabolism of proteins and nucleic acids and in mitotic cell division.

Zinc tends to be less toxic than other heavy metals, and the difference between essential and toxic levels is large.[35] There is evidence that excess zinc in blood is bound by albumin.[140] Symptoms of zinc toxicity are vomiting, dehydration, electrolyte imbalance, stomach pain, nausea, lethargy, dizziness, and muscular incoordination. The emetic dose of zinc in water[141] is 675 to 2280 mg/ℓ. A case of zinc poisoning quoted by Prasad[142] involved a patient with renal failure who underwent hemodialysis against water which had been stored in a zinc galvanized tank. The patient exhibited nausea, vomiting, fever, and severe anemia. An epidemiological study in South Wales reported no significant correlation between CNS malformation and zinc concentrations in drinking water.[143] A community water supply survey carried out in the U.S. in 1970 found that the most widespread element in drinking water was zinc, for which a 2-ℓ consumption would give an average intake[144] of 390 µg/day, which is considerably less than the daily dietary allowance for adult males of 11 mg recommended[93] by the WHO.

The role of zinc as a carcinogen or cocarcinogen is unclear. Hypertension induced by cadmium can be reduced by zinc,[145] and a strong positive correlation of cadmium and zinc in blood samples from New Jersey children has been observed. In animal studies, the toxicity of cadmium in drinking water was found to be less at a molar Cd/Zn ratio of 1:4 than 1:1.[146]

I. Copper

Copper is an essential element used in processes of blood formation and iron utilization, with a human daily requirement of 0.03 mg/kg for adults. The main intake is normally through the diet, although the proportion contributed by drinking water varies with water hardness through corrosion of domestic piping. In Canada, drinking water in hard-water areas provided 0.6 mg of copper per person, but 1.2 mg in soft-water areas.[92] Maximum concentrations in drinking water in a survey of 100 American cities were 1.67 mg/ℓ in running water and 2.41 mg/ℓ in standing water;[144] these concentrations exceed the EPA recommended limit of 1 mg/ℓ in domestic water supplies. Underwood[11] indicated that such levels may be hazardous, particularly in areas where iron and zinc intake are low, because of the metabolic antagonism between copper and these metals.

Copper is absorbed through the intestinal tract (40 to 70%) although, like zinc, absorption varies with intake. Its half-life in humans is 80 days and the total body burden is 75 to 150 mg.[147] Once absorbed, copper is transported in the blood and stored normally in muscle, liver, and brain tissue. In cases of acute toxicity, copper is found in the brain, liver, stomach, hair roots, and urine, with the hair levels reflecting copper status at the time of growth. Symptoms of acute toxicity are gastric ulcers, hemolysis, jaundice, hepatic necrosis, and renal damage. Chronic toxicity in man is found infrequently but has been reported to cause "pink disease" in infants ingesting copper-contaminated water at concentrations[54] of 0.8 mg/ℓ. Various studies cited by Demayo et al.[148] reported that elevated levels of copper are associated with disorders such as biliary atresia and cirrhosis in children, Hodgkin's disease, leukemia, atherosclerosis, and hypertension. Copper has been shown to be carcinogenic to animals, but not conclusively to humans.[72]

J. Aluminum

Aluminum is one of the most widely distributed metals, but despite relatively consistent exposure to high levels, the human burden tends to remain constant at approximately 30 mg.[70] This has been attributed to its poor absorbance through the GI tract. Exposure via other routes has, however, been shown to be more hazardous, as evidenced by the occurrence of dialysis encephalophathy and osteomalacia in renal dialysis patients. These diseases have been ascribed to the high concentrations of aluminum in dialysis media originating from the use of aluminum as a coagulent in water purification.[149] Aluminum has further been implicated in brain disease[150] and behavioral disorders in humans.[151] The health effects of aluminum on mammals have been reviewed by Krueger et al.[152] Animal studies have indicated that the deleterious effects of aluminum vary with zinc and copper levels in the body, possibly through competition for biological binding sites.[153]

K. Beryllium

Beryllium exerts a toxic effect by competing with magnesium for enzyme sites and inhibiting DNA polymerase, thymidine kinase, and alkaline phosphatase.[46] Most documented human health effects have been due to airborne beryllium or direct contact, such as berylliosis by accumulation in the lung, damage to skin and mucous membrane, and effects on allergic sensitization and immunological reactions.[154] Epidemiological data suggest it to be weakly carcinogenic; however, despite its cumulative nature, there is evidence that an immunity may be acquired over long periods of human exposure.[154]

L. Thallium

Thallium occurs in monovalent and trivalent forms, both of which are toxic. In mammals, the range of LD_{50} values of thallium(I) inorganic species has been quoted as 16 to 71 mg/kg and for thallium(III) as 6 to 74 mg/kg; for humans, doses of above 14 mg/kg are fatal.[48] Absorption of thallium is complete via any route of exposure, following which the metal is rapidly distributed in the intracellular space of most tissues. The biological life is 3 to 8 days.[155]

Symptoms of acute oral exposure are normally delayed for 1 to 2 days, then manifested as GI damage, limb pain and paralysis, polyneuritis, raised blood pressure, blindness, liver and kidney damage, and delayed hair loss. The systemic effects are frequently not attributed to thallium poisoning until hair loss begins to occur; a variety of diagnoses are typical for victims murdered by thallium poisoning.

Thallium(I) exerts its toxic effects by uncoupling mitochondrial oxidative phosphorylation, inhibiting enzymes such as alkaline phosphatase, interfering with protein synthesis and ribosome aggregation, and causing antimitotic effects. There is evidence from animal studies that selenium and sulfur have an antagonistic influence on thallium and can protect against its toxicity.[48]

From their extensive review of the literature, Smith and Carson[48] reported that there is little evidence of human intoxication arising from environmental thallium. Background levels of thallium in water are generally of the order of 2 µg/ℓ, but thallium-polluted water may provide a higher contribution to intake than food; levels of 10 to 20 µg/day may be ingested and cause chronic toxic effects. Where abnormally high thallium levels occur, however, as in leachates from sulfide ore processing or metal smelting wastes, its effects may be difficult to isolate since other metals are also likely to be present at elevated concentrations.

M. Silver

Data cited by Smith and Carson[156] suggest that the GI tract is the most important uptake route for silver, and that the rate of absorption is species dependent. Silver chloride, such as may be produced during chlorination in water treatment, is absorbed more rapidly than colloidal silver species found in natural waters and sewage effluents. Although absorption in animals is low (10%) and retention even lower (3 to 10%), silver can be deposited in all tissues and is not liberated. The human body burden is approximately 10 mg/kg, but in very extreme cases, it has been recorded as 1300 mg/kg. Acute toxic effects include necrosis, hemorrhage, GI symptoms, and pulmonary edema, while chronic toxicity is manifested by argyria, a grey pigmentation of the skin and hair resulting from accumulation of silver in the tissues. Most recorded cases of silver poisoning have resulted from accidental or suicidal overdoses of medical forms of silver; Smith and Carson[156] concluded that toxicity arising from environmental silver was unlikely to pose a significant threat, due to the poor absorption and retention of silver by humans.

N. Tellurium

Tellurium is normally ingested in much higher concentrations from the diet than from water. The largest accumulation in humans is found in bone, and a typical body burden of 600 mg has been reported.[157] The species of tellurium to which an individual is exposed affects the degree of absorption and consequent toxicity. Elemental tellurium ingested orally is poorly absorbed from the intestinal tract, whereas sodium tellurite is 25 to 50% absorbed. Tellurites are ten times as toxic as tellurates. Once absorbed, soluble tellurites are reduced to tellurides and then methylated; the subsequent exhalation of dimethyl telluride gives a strong, characteristic, garlicky odor to the breath.[157] Although tellurium crosses the placenta and blood-brain barrier in animals (the latter resulting in damage through the precipitation of crystalline tellurium within the brain), no embryotoxic or teratogenic effects have been recorded in humans.[72]

O. Vanadium

In solution, vanadium forms inorganic orthovanadate and oxyanions, which are poorly absorbed from the GI tract. Absorption of vanadium differs depending on the mode of exposure; the toxicity of orally administered vanadium to animals is generally less than that given by injection or inhalation. Vanadate concentrates in the kidneys, and hence urinary elimination provides a good indication of exposure.[158] The half-life in humans in 15 to 20 hr, and elimination curves observed in exposed workers led Sabboni and Maroni[159] to conclude that long-term vanadium accumulation was likely during low-level, continuous exposure.

The toxicological mode of action of vanadium may be related in part to its inhibitory effect on Na^+, K^+-adenosinetriphosphatase[160] and it has additionally been reported to interfere in the synthesis and activity of coenzymes A and Q, monamine oxidase, cholesterol, and phospholipids.[161] Manifestations of toxicity from airborne vanadium include irritative effects on upper-airway mucous membranes and acute respiratory failure, with a green tongue coloration characteristically observed. Epidemiological studies of populations in the U.K.[162] and U.S.[163] correlated ambient air concentrations of vanadium with various forms of heart disease and cancer. However, few such studies appear to have been made on the potential health effects of vanadium in water.

IV. REASONS FOR CONCERN

A fundamental factor which heightens the concern over the presence of potentially toxic heavy metals in the environment is their nonbiodegradability and consequent persistence. This is further magnified by the tendency of some metals, notably mercury and cadmium, to become concentrated in food chains through bioaccumulation. One of the first groups of organisms to be affected by metal-laden discharges to the aquatic environment are bacteria, certain species of which have been observed to exhibit plasmid-determined resistance to heavy metals. Strains resistant to antimony, arsenic, boron, cadmium, chromium, cobalt, copper, mercury, lead, nickel, silver, tellurium, and zinc have been reported and the use of such genetically modified bacteria has been proposed for the treatment of heavily metal-contaminated wastewaters.[164,165] However, resistance in higher organisms is less easily acquired, and metals may be accumulated to lethal or sublethal levels in fish through a food chain such as sediment → bacteria → tubificids → fish.[166] Some life forms are particularly adept at accumulating metals; for instance, drops of elemental mercury have been found in chickweed leaves.[46] In some cases, as exemplified by the Minimata Bay episode, bioaccumulation via the food chain extends upwards to man.

In addition to catastrophic incidences of large-scale localized inputs of metals from industrial sources, problems arise through gradual increases in metal concentrations from nonpoint sources, particularly where water reuse is practiced. Approximately one third of water supplies in the U.K. are obtained from lowland rivers, which simultaneously serve to transport domestic and industrial wastewaters to sea. Wastewater may constitute an average of 2 to 20% of the total flow of some British rivers, with a maximum of >100% of the flow of the lower Thames during dry weather conditions.[167] Accumulation of nonbiodegradable mineral content is inherent in water reuse due to the nature of wastewater treatment processes,[168] and, as it seems that the practice of water reuse is likely to increase in the future, this aspect constitutes considerable cause for concern with regard to public health. Analytical, epidemiological, and toxicological studies are currently being undertaken to investigate the possible health hazards arising from water reuse, including the incidence of cancer of the GI and genitourinary tracts.[167]

Concomitant with the problems posed by dispersal of liquid effluents are those of sludge disposal, discussed further in Volume II, Chapters 5 and 6. Wastewater treatment in the U.K. produces some 80,000 tonnes of wet sludge per day, of which approximately 40% is ultimately used in agriculture and 30% disposed of to sea.[169] Metals removed from the soluble phase during wastewater treatment may be prevented from entering clean water bodies via works effluent, but nevertheless still have to be disposed of in the resultant sludge without remobilization or translocation occurring. Hazards arising from the disposal of sludge to land incorporate phytotoxicity (detrimental to plant health and growth) and zootoxicity (the intoxication of animals through consumption of contaminated crops). The latter is more insidious since there are no outward indicators of toxicity exhibited by plants at levels likely to affect either animal or human (direct or indirect) food chains.[2]

The potential for toxic effects on humans resulting from sludge disposal to land varies with a multiplicity of factors, such as the nature and metal content of the soil and the sludge, the rate and frequency of sludge application, the species of metals in the sludge and subsequently in the sludge-amended soil, the type of crop grown, and the part of the plant consumed. As an example of the latter, zinc and notably cadmium are absorbed easily by plant roots and translocated to upper portions, particularly in leafy vegetables like lettuce, while copper and lead are retained in the roots. Of the most hazardous metals, some are relatively immobile in the soil system (e.g., lead), while others, such as mercury and arsenic, are rapidly converted to unavailable forms. Yet others, such as molybdenum and selenium, are unlikely to be present in sewage sludges in significant amounts. The most critical of sludge-associated metals, in terms of potential toxicity, is cadmium, which is readily cumulative. Evidence from a Ministry of Agriculture, Fisheries, and Foods (MAFF) dietary survey[170] showed that people consuming vegetables grown on sludged land contaminated with cadmium had higher than normal cadmium intakes, although none of them suffered from heavy-metal-related diseases. Davis et al.,[171] however, recently calculated that current U.K. guidelines for sludge disposal to land preclude the possibility of an individual exceeding the WHO acceptable daily cadmium intake of 70 μg/day, even if all the crops consumed originated from sludged soils. However, plant uptake is not the only route of metal accumulation, and a further hazard resulting from sludge disposal to land is the potential for its direct ingestion by grazing animals and subsequent accumulation of metal content in the kidneys and liver.[172]

Disposal of sludge to sea is restricted by the 1974 Dumping at Sea Act to licensed sites generally chosen to minimize the environmental impact; however, the rapid dispersion of sludge at sea precludes such close monitoring of localized effects as is possible on land areas used for sludge disposal. The great propensity for fish, and shellfish in particular, to accumulate and concentrate heavy metals has led to their utilization as biomonitors and indicators of sites of contamination. As parts of marine food chains, filter-feeding bivalve mollusks can accumulate high concentrations of metals; Bryan[173] recorded a concentration factor of 4 million for zinc in mollusk kidney tissue. Thus potential hazards to man through consumption of food items originating from the marine environment exist.

Concern over the discharge and disposal of heavy metals encompasses not only their deleterious effects towards man but also to the receiving environment. The environmental risk posed by metals to aquatic systems can be divided into hazards pertaining to terrestrial organisms utilizing the aquatic system, as previously discussed, and those relating to the components of the aquatic ecosystem itself. The latter comprises the extinction of plant or animal species, loss of species from specific ecosystems, alterations in species diversity and abundance, alterations in population characteristics (e.g., age structure, individuals' size, standing crop), interference in functions of energy conversion and element recycling, and alterations in the physiochemical properties of the system (e.g., appearance, odor). The degree or intensity, the geographical extent, and the duration of the harmful effects caused by metal contamination can be related to the properties of the pollutant source and the intrinsic properties of the receiving environment. A summary of these determinants of heavy metal pollution is given in Table 4.

Table 4
PROPERTIES OF SOURCE, SUBSTANCE, AND ENVIRONMENT INFLUENCING THE EFFECTS EXERTED BY HEAVY METALS IN AQUATIC SYSTEMS[174]

Source properties
   Location, magnitude and time course of input
   Physicochemical species of input material

Substance properties
   Limit and rate of solution/precipitation
   Partitioning behavior between components of aquatic system
   Disposition within organisms and resultant toxicity
   Persistence and bioaccumulation potential
   Antagonistic/synergistic potential
   Susceptibility to mobilization and redistribution

Environmental properties
   Volume of receiving water body
   Water reuse and renewal rates
   Physicochemical properties of aquatic system
   Type and trophic status of biota present
   Nature of species assemblages and foodchains

REFERENCES

1. **Nieboer, E. and Richardson, D. H. S.**, The replacement of the nondescript term "heavy metals" by a biologically and chemically significant classification of metal ions, *Environ. Pollut. Ser. B.*, 1, 3, 1980.
2. **Matthews, P. J.**, Control of metal application rates from sewage sludge utilization in agriculture, *Crit. Rev. Environ. Control*, 14, 199, 1984.
3. **Boudene, C.**, Food contamination by metals, in *Trace Metals: Exposure and Health Effects*, Di Ferrante, E., Ed., Pergamon Press, Oxford, 1979, 163.
4. **Jarvis, P. J.**, *Heavy Metal Pollution: An Annotated Bibliography*, Geo Books, Norwich, U.K., 1983.
5. **Pearson, R. G.**, Hard and soft acids and bases, HSAB. I. Fundamental principles, *J. Chem. Educ.*, 45, 581, 1968.
6. **Förstner, U. and Wittman, G.**, *Metal Pollution in the Aquatic Environment*, Springer-Verlag, Berlin, 1979, chap. B.
7. **Barth, E. F.**, Discussion of "The effects and removal of heavy metals in biological treatment", by **Adams, C.** in *Heavy Metals in the Aquatic Environment*, Krenkel, P. A., Ed., Pergamon Press, Oxford, 1973, 293.
8. **Truhaut, R.**, Interaction of metals, in *Trace Metals: Exposure and Health Effects*, Di Ferrante, E., Ed., Pergamon Press, Oxford, 1979, 147.
9. **Duffus, J. H.**, *Environmental Toxicology*, Edward Arnold, London, 1980, chap. 1.
10. **Shacklette, H. T., Erdman, J. A., Harms, T. F., and Papp, C. S. E.**, Trace elements in plant foodstuffs, in *Toxicity of Heavy Metals in the Environment*, Oehme, F. W., Ed., Marcel Dekker, New York, 1978, 25.
11. **Underwood, E. J.**, Environmental sources of heavy metals and their toxicity to man and animals, *Prog. Water Technol.*, 11, 33, 1979.
12. **Florence, T. M.**, The speciation of trace elements in water, *Talanta*, 29, 345, 1982.
13. **Wood, J. M.**, Biological cycles for toxic elements in the environment, *Science*, 183, 1049, 1974.
14. **Pearson, R. G.**, Hard and soft acids and bases, HSAB. II. Underlying theories, *J. Chem. Educ.*, 45, 643, 1968.
15. **Ewan, R. C.**, Toxicology and adverse effects of mineral imbalance with emphasis on selenium and other minerals, in *Toxicity of Heavy Metals in the Environment*, Oehme, F. W., Ed., Marcel Dekker, New York, 1978, 445.
16. **Florence, T. M.**, Trace element speciation and aquatic toxicity, *Trends Anal. Chem.*, 2, 162, 1983.

17. **Nelson, A. and Donkin, P.**, Processes of bioaccumulation; the importance of chemical speciation, *Mar. Pollut. Bull.*, 16, 164, 1985.
18. **Stumm, W.**, Metal pollutants in waters: their effect, controlling factors and ultimate fate, in *Proc. Int. Conf. Heavy Metals in the Environment*, CEP Consultants, Edinburgh, 1983, 1.
19. **Morel, F. M. M., McDuff, R. G., and Morgan, J. J.**, Interactions and chemostasis in aquatic chemical systems: role of pH, pE. Solubility and complexation, in *Trace Metals and Metal Organic Interactions in Natural Waters*, Ann Arbor Science, Ann Arbor, Mich., 1973, chap. 6.
20. **Richter, R. O. and Theis, T. L.**, Nickel speciation in a soil water system, in *Nickel in the Environment*, Nriagu, J. O., Ed., John Wiley & Sons, New York, 1980, chap. 2.
21. **Perhac, R. M.**, Distribution of Cd, Co, Cu, Fe, Mn, Ni, Pb, Zr and Zn in dissolved and particulate solids from two streams in Tennessee, *J. Hydrol.*, 15, 177, 1972.
22. **Gibbs, R. J.**, Transport phases of transition metals in the Amazon and Yukon rivers, *Geol. Soc. Am. Bull.*, 88, 829, 1977.
23. **Snodgrass, W. J.**, Distribution and behaviour of nickel in the aquatic environment, in *Nickel in the Environment*, Nriagu, J. O., Ed., John Wiley & Sons, New York, 1980, chap. 9.
24. **Mance, G. and Yates, J.**, Proposed Environmental Quality Standards for List II Substances in Water: Nickel, Technical Report TR211, Water Research Centre, Medmenham, U.K., 1984.
25. **Rudd, T., Sterritt, R. M., and Lester, J. N.**, Formation and conditional stability constants of complexes formed between heavy metals and bacterial extracellular polymers, *Water Res.*, 18, 379, 1984.
26. **Sillen, L. G. and Martell, A. E.**, Stability constants of metal ion complexes, *Spec. Publ. Chem. Soc. London*, 17, 1964.
27. **Alabaster, J. S. and Lloyd, R.**, *Water Quality Criteria for Freshwater Fish*, Butterworths, London, 1980, chap. 9.
28. **Guy, R. D. and Chakrabarti, C. L.**, Studies of metal-organic interactions in model systems pertaining to natural water systems, *Can. J. Chem.*, 54, 2600, 1976.
29. **Mance, G., Brown, V. M., and Yates, J.**, Proposed Environmental Quality Standards for List II Substances in Water: Copper, Technical Report TR210, Water Research Centre, Medmenham, U.K., 1984.
30. **Gardiner, J.**, The chemistry of cadmium in natural water. I. A study of cadmium complex formation using the cadmium specific-ion electrode, *Water Res.*, 8, 23, 1974.
31. **Alabaster, J. S. and Lloyd, R.**, *Water Quality Criteria for Freshwater Fish*, Butterworths, London, 1980, chap. 10.
32. **Sterritt, R. M. and Lester, J. N.**, The influence of sludge age on heavy metal removal in the activated sludge process, *Water Res.*, 15, 59, 1981.
33. **Stephenson, T. and Lester, J. N.**, Heavy metal behavior during the activated sludge process. I. Extent of soluble and insoluble metal removal, *Sci. Total Environ.*, 63, 199, 1987.
34. **Mance, G. and Yates, J.**, Proposed Environmental Quality Standards for List II Substances in Water: Zinc, Technical Report TR 209, Water Research Centre, Medmenham, U.K., 1984.
35. **Taylor, M. C., Demayo, A., and Taylor, K. W.**, Effects of zinc on humans, laboratory and farm animals, terrestrial plants and freshwater aquatic life, *Crit. Rev. Environ. Control.*, 12, 133, 1982.
36. **Demayo, A., Taylor, M. C., Taylor, K. W., and Hodson, P. V.**, Toxic effects of lead and lead compounds on human health, aquatic life, wildlife, plants and livestock, *Crit. Rev. Environ. Control*, 12, 257, 1982.
37. **Brown, V. M., Gardiner, J., and Yates, J.**, Proposed Environmental Quality Standards for List II substances in Water: Inorganic Lead, Technical Report TR 208, Water Research Centre, Medmenham, U.K., 1984.
38. **Stoveland, S. and Lester, J. N.**, A study of the factors which influence metal removal in the activated sludge process, *Sci. Total Environ.*, 16, 37, 1980.
39. **Train, R. E.**, Quality Criteria for Water, U.S. Environmental Protection Agency, Washington, D.C., 1979.
40. **Schwartz, M. K.**, Role of trace elements in cancer, *Cancer Res.*, 35, 3481, 1975.
41. **The Open University**, *Second Level Chemistry Data Book*, Open University Press, Milton Keynes, U.K., 1976, 19.
42. **Wong, P. T. S., Chau, Y. K., and Luxon, P. L.**, Methylation of lead in the environment, *Nature (London)*, 253, 263, 1975.
43. **Huey, C. W., Brinkman, F. E., Iverson, W. P., and Grim, S. O.**, Bacterial volatilization of cadmium, Proc. Int. Conf. Heavy Metals in the Environment, Toronto, 1975, 214.
44. **Fleming, R. W. and Alexander, M.**, Dimethyl selenide and dimethyl telluride formation by a strain of Penicillium, *Appl. Microbiol.*, 24, 424, 1972.
45. **Cox, D. P. and Alexander, M.**, Production of trimethylarsine gas from various compounds by three sewage fungi, *Bull. Environ. Contam. Toxicol.*, 9, 84, 1973.
46. **Duffus, J. H.**, *Environmental Toxicology*, Edward Arnold, London, 1980, chap. 6.
47. **Brown, H. G., Hensley, C. P., McKinney, G. L., and Robinson, J. L.**, Efficiency of heavy metals removal in municipal sewage treatment plants, *Environ. Lett.*, 5, 103, 1973.

48. **Smith, I. C. and Carson, B. L.**, *Trace Metals in the Environment, Vol. 1, Thallium,* Ann Arbor Science, Ann Arbor, Mich., 1977.
49. **Jackson, T. A.**, Relationships between the properties of heavy metals and their biogeochemical behaviour in Lakes and river-lake systems, in *Proc. Int. Conf. Heavy Metals in the Environment,* CEP Consultants, Edinburgh, 1979, 457.
50. **Steinberg, C.**, Species of dissolved metals derived from oligotrophic hard water, *Water Res.,* 14, 1239, 1980.
51. **Lester, J. N.**, Significance and behaviour of heavy metals in waste water treatment processes. I. Sewage treatment and effluent discharge, *Sci. Total Environ.,* 30, 1, 1983.
52. **Bloomfield, C. and McGrath, S. P.**, A comparison of the extractabilities of Zn, Cu, Ni and Cd from sewage sludges prepared by treating raw sewage with the metal salts before and after anaerobic digestion, *Environ. Pollut. Ser. B.,* 3, 193, 1982.
53. **Legret, M., Demare, D., and Marchandise, P.**, Speciation of heavy metals in sewage sludges, *Proc. Int. Conf. Heavy Metals in the Environment,* CEP Consultants, Edinburgh, 1983, 350.
54. National Academy of Sciences, *Drinking Water and Health,* Vol. I, National Academy of Sciences, Washington, D.C., 1977.
55. **Luckey, T. D. and Venugopal, B.**, *Metal Toxicity in Mammals,* Vol. 1, Plenum Press, New York, 1977.
56. **Campbell, P. G. C., Stokes, P. M., and Galloway, J. N.**, Effects of atmospheric deposition on the geochemical cycling and biological availability of metals, in *Proc. Int. Conf. Heavy Metals in the Environment,* CEP Consultants, Edinburgh, 1983, 760.
57. **Faucherre, J., Pinart, A. M., Pinart, J., and Dutot, A.**, A regional case study of the pollution of natural waters, soils and plants by lead, cadmium and zinc, in *Pollutants and their Ecotoxicological Significance,* Nurnberg, H. W., Ed., John Wiley & Sons, Chichester, 1985, chap. 14.
58. **Walker, C.**, *Environmental Pollution by Chemicals,* 2nd ed., Hutchinson & Co., London, 1975, chap. 4.
59. **Garattini, S.**, Difficulties in extrapolating toxicological data from animals to man, in *Safety Evaluation and Regulation of Chemicals,* Homburger, F., Ed., S. Karger, Basel, 1982, 85.
60. **Williamson, S. J.**, Epidemiological studies on cancer and organic compounds in U.S. drinking waters, *Sci. Total Environ.,* 18, 187, 1981.
61. **Kobayashi, J.**, Relation between 'Itai itai' disease and the pollution of river water by cadmium from a mine, in *Int. Conf. of Advances in Water Pollution Research,* Vol. 1, Jenkins, S. H., Ed., Pergamon Press, Oxford, 1970, 1.
62. **Harada, M.**, Methyl mercury poisoning due to environmental contamination ('Minimata Disease'), in *Toxicity of Heavy Metals in the Environment,* Oehme, F. W., Ed., Marcel Dekker, New York, 1978, 261.
63. **Smeets, J. and Amavis, R.**, European communities directive relating to the quality of water intended for human consumption, *Water Air Soil Pollut.,* 15, 483, 1982.
64. United States Environmental Protection Agency, National revised primary drinking water regulations; advance notice of proposed rulemaking, *Fed. Regist.,* 48, 45502, 1983.
65. **Buck, W. B.**, Copper/molybdenum toxicity in animals, in *Toxicity of Heavy Metals in the Environment,* Oehme, F., Ed., Marcel Dekker, New York, 1978, 491.
66. **Olguin, A., Jange, P., Cebrian, M., and Albores, A.**, Arsenic levels in blood, urine, hair and nails from a chronically exposed population, *Proc. West. Pharmacol. Soc.,* 26, 175, 1983.
67. **Lafontaine, A.**, Health effects of arsenic, in *Trace Metals: Exposure and Health Effects,* Di Ferrante, E., Ed., Pergamon Press, Oxford, 1979, 107.
68. **Beaudoin, A. R.**, Teratogenicity of sodium arsenite in rats, *Teratology,* 10, 153, 1974.
69. **Léonard, A.**, The mutagenic and genotoxic effects of heavy metals, in *Proc. Int. Conf. Heavy Metals in the Environment,* CEP Consultants, Edinburgh, 1983, 700.
70. **Becking, G. C.**, Recent advances in the toxicity of heavy metals — an overview, *Fundam. Appl. Toxicol.,* 1, 348, 1981.
71. **Frost, D. V.**, What do losses in selenium and arsenic bioavailability signify for health?, *Sci. Total Environ.,* 28, 455, 1983.
72. **Kurzel, R. B. and Cetrulo, C. L.**, The effect of environmental pollutants on human reproduction, including birth defects, *Environ. Sci. Technol.,* 15, 626, 1981.
73. **Lugo, G., Cassady, G., and Palmisano, P.**, Acute maternal arsenic intoxication with neonatal death, *Am. J. Dis. Child.,* 117, 328, 1969.
74. World Health Organization, European Standards for Drinking Water, 2nd ed., World Health Organization, Geneva, 1970.
75. World Health Organization, International Standards for Drinking Water, 3rd ed., World Health Organization, Geneva, 1971.
76. U.S. Environmental Protection Agency, Water quality criteria documents; availability, *Fed. Regist.,* 45, 79318, 1980.
77. World Health Organization, WHO Environmental health criteria for cadmium, *Ambio,* 6, 287, 1977.

78. Commission of the European Communities, *Criteria (Dose/Effect Relationships) for Cadmium,* Pergamon Press, Oxford, 1978.
79. **Nriagu, J. O.,** Ed., *Cadmium in the Environment, Part 2, Health Effects,* John Wiley & Sons, New York, 1980.
80. **Fielder, R. J., Dale, E. A., Sorrie, G. S., Bishop, C. M., Van den Heuvel, M. J., Pryde, E., and Fletcher, A. P.,** Cadmium and its compounds, *Toxicol. Rev.,* 7, 1, 1983.
81. **Rahola, T., Aaran, R. K., and Miettinen, J. K.,** Half-time studies of mercury and cadmium by whole body counting, in *Assessment of Radioactive Contamination in Man,* IAEA, Vienna, 1972.
82. **Lauwerys, R. R.,** Health effects of cadmium, in *Trace Metals: Exposure and Health Effects,* Di Ferrente, E., Ed., Pergamon Press, Oxford, 1979, 43.
83. **Nordberg, G. F.,** Factors influencing metabolism and toxicity of metals, in *Trace Metals: Exposure and Health Effects,* Di Ferrante, E., Ed., Pergamon Press, Oxford, 1979, 157.
84. **Ragan, H. A.,** The bioavailability of iron, lead, and cadmium via gastrointestinal adsorption: a review, *Sci. Total Environ.,* 28, 317, 1983.
85. **Friberg, L., Piscator, M., Nordberg, G. F., and Kjelleström, T.,** *Cadmium in the Environment,* 2nd ed., CRC Press, Cleveland, 1974.
86. **Piscator, M.,** Exposure to cadmium, in *Trace Metals: Exposure and Health Effects,* Di Ferrante, E., Ed., Pergamon Press, Oxford, 1979, 35.
87. **Vivoli, G., Bergomi, M., Borella, P., Fantuzzi, G., and Caselgrandi, E.,** Interaction between cadmium and some biochemical parameters involved with human hypertension, in *Proc. Int. Conf. Heavy Metals in the Environment,* CEP Consultants, Edinburgh, 1983, 545.
88. **Tsuchiya, H., Mitani, K., Kodama, K., and Nakata, T.,** Placental transfer of heavy metals in normal pregnant Japanese women, *Arch. Environ. Health,* 39, 11, 1984.
89. World Health Organization, Lead, Environmental Health Criteria 3, World Health Organization, Geneva, 1977.
90. Department of Health and Social Security, Lead and Health, Report of a DHSS Working Party on lead in the environment, Her Majesty's Stationery Office, London, 1980.
91. Royal Commission on Environmental Pollution, Lead in the Environment, 9th Report, Cmnd. 8852, Her Majesty's Stationary Office, London, 1983.
92. **Neri, L. C.,** Some data from Canada, in *Hardness of Drinking Water and Public Health,* Proc. Eur. Sci. Coll, Luxembourg, Commission of the European Communities, Pergamon Press, New York, 1975, 343.
93. World Health Organization, The Hazards to Health of Persistent Substances in Water, World Health Organization, Copenhagen, 1973.
94. **Barry, P. S. I.,** A comparison of concentrations of lead in human tissues, *Br. J. Ind. Med.,* 32, 119, 1975.
95. **Goyer, R. A. and Mushak, P.,** Lead toxicity, laboratory aspects, *Adv. Mod. Toxicol.,* 2, 41, 1977.
96. **Yule, W. and Lansdown, R.,** Lead and children's development: recent findings, in *Proc. Int. Conf. Heavy Metals in the Environment,* CEP Consultants, Edinburgh, 1983, 912.
97. **Wilson, A. T.,** Effects of abnormal lead content of water supplies on maternity patients. The use of a simple industrial screening test in ante-natal care in general practice, *Scott. Med. J.,* 11, 73, 1966.
98. **Bonnefoy, X., Huel, G., and Guéguen, R.,** Variation de la plombémie en fonction de la contamination par le plomb de l'eau livrée à la consommation, *Water Res.,* 19, 1299, 1985.
99. **Quinn, M. J.,** Factors affecting blood lead concentrations in the U.K., in *Proc. Int. Conf. Heavy Metals in the Environment,* CEP Consultants, Edinburgh, 1983, 294.
100. National Research Council, Drinking Water and Health, Vol. 4, National Research Council, Washington, D.C., 1982.
101. World Health Organization, Guidelines for Drinking Water Quality, Vol. 1 (Draft), World Health Organization Regional Office for Europe, Copenhagen, 1982.
102. WHO Expert Committee on Food Additives, Evaluation of Certain Food Additives and the Contaminants Mercury, Lead and Cadmium, World Health Organization Tech. Rep. Ser. No. 505, Geneva, 1972.
103. **Gatti, G. L., Macri, A., and Silano, V.,** Biological and health effects of mercury, in *Trace Metals: Exposure and Health Effects,* Di Ferrante, E., Ed., Pergamon Press, Oxford, 1979, 73.
104. **Suzuki, T., Shishido, S., and Ishihara, N.,** Interaction of inorganic to organic mercury in their metabolism in human body, *Int. Arch. Occup. Environ. Health,* 38, 103, 1976.
105. **Buncel, E., Kumar, R., Norris, A. R., Racz, W. J., Taylor, S. E., and Vanderwater, L. J. S.,** Metal ion biomolecule interactions — binding of methylmercury to nucleosides and neurotoxicological effects, in *Proc. Int. Conf. Heavy Metals in the Environment,* CEP Consultants, Edinburgh, 1983, 609.
106. **Kosta, L., Byrue, A. R., and Zelenko, V.,** Correlation between selenium and mercury in man following exposure to inorganic mercury, *Nature (London),* 254, 238, 1975.
107. **Triebig, G. and Schaller, K. H.,** Relationship between mercury levels in blood and urine and nerve conduction velocities, in *Proc. Int. Conf. Heavy Metals in the Environment,* CEP Consultants, Edinburgh, 1981, 470.

108. World Health Organization, Mercury, Environmental Health Criteria 1, World Health Organization, Geneva, 1976.
109. **Magos, L.,** Health effects of mercury, in *Trace Metals: Exposure and Health Effects*, Di Ferrante, E., Ed., Pergamon Press, Oxford, 1979, 221.
110. **Sunderman, F. W.,** A review of the metabolism and toxicology of nickel, *Ann. Clin. Lab. Sci.*, 7, 377, 1977.
111. **Nielsen, F. H.,** Nickel toxicity, in *Advances in Modern Toxicology, Vol. 2, Toxicology of Trace Elements*, Goyer, R. A. and Mehlam, M. A., Eds., Hemisphere, London, 1977.
112. **Mushak, P.,** Metabolism and systemic toxicity of nickel, in *Nickel in the Environment*, Nriagu, J. O., Ed., John Wiley & Sons, New York, 1980, chap. 20.
113. **Bencko, V.,** Nickel: a review of its occupational and environmental toxicology, *J. Hyg. Epidemiol. Microbiol. Immunol.*, 27, 237, 1983.
114. **Norseth, T.,** Health effects of nickel and chromium, in *Trace Metals: Exposure and Health Effects*, Di Ferrante, E., Ed., Pergamon Press, Oxford, 1979, 135.
115. **Ho, W. and Furst, A.,** Nickel excretion by rats following a single treatment, *Proc. West. Pharmacol. Soc.*, 16, 245, 1973.
116. Committee on Medical and Biological Effects of Environmental Pollutants, Nickel, National Academy of Sciences, Washington, D.C., 1975.
117. **Ung, J. and Furst, A.,** Solubility of nickel powder in human serum, *Sci. Total Environ.*, 28, 399, 1983.
118. **Sevin, I. F.,** Nickel, in *Metals in the Environment*, Waldron, H. A., Ed., Academic Press, London, 1980.
119. **Sunderman, F. W.,** Nickel poisoning, in *Laboratory Diagnosis of Diseases caused by Toxic Agents*, Sunderman, F. W. and Sunderman, F. W., Jr., Eds., W. H. Green, St. Louis, Mo., 1970, 387.
120. **Spruitt, D. and Bongaarts, P. J. M.,** Nickel content of plasma, urine and hair in contact dermatitis, *Dermatologica*, 154, 291, 1977.
121. **Furst, A. and Radding, S. B.,** An update on nickel carcinogenesis, in *Nickel in the Environment*, Nriagu, J. O. Ed., John Wiley & Sons, New York, 1980, chap. 24.
122. Council of the European Communities, Council directive relating to the quality of water intended for human consumption (80/778/EEC), *Off. J. Eur. Commun.*, L229, 11, 1980.
123. **Mertz, W.,** Effects and metabolism of glucose tolerance factor, *Nutr. Rev.*, 33, 129, 1975.
124. **Rondia, D.,** Sources, modes and levels of human exposure to chromium and nickel, in *Trace Metals: Exposure and Health Effects*, Di Ferrante, E., Ed., Pergamon Press, Oxford, 1979, 117.
125. **Langard, S.,** Chromium, in *Metals in the Environment*, Waldron, H. A., Ed., Academic Press, London, 1980.
126. **Beyersman, D., Köster, A., Buttner, B., and Flessel, P.,** Chromium uptake and toxicity with mammalian and bacterial cells, in *Proc. Int. Conf. Heavy Metals in the Environment*, CEP Consultants, Edinburgh, 1983, 537.
127. **Beretic, T., Kovac, S., and Bogadi, A.,** Toxic, allergic and carcinogenic effects of chromium in electroplating and welding, in *Proc. Int. Conf. Heavy Metals in the Environment*, CEP Consultants, Edinburgh, 1983, 507.
128. **Schroeder, D. C. and Lee, G. F.,** Potential transformations of chromium in natural waters, *Water Air Soil Pollut.*, 4, 355, 1975.
129. National Academy of Sciences, Drinking Water and Health, Vol. 3, National Academy of Sciences, Washington, D.C., 1980.
130. Committee on Medical and Biological Effects of Environmental Pollutants, Selenium, Medical and Biological Effects, National Academy of Sciences, Washington, D.C., 1976.
131. **Spallholz, J. E., Martin, J. L., and Gunther, H. E., Eds.,** *Selenium in Biology and Medicine*, Proc. 2nd Int. Symp. Selenium, AVI, Westport, Conn., 1981.
132. **Rotruck, J. T., Pope, A. L., Gauther, H. E., Swanson, A. B., Hafeman, D. G., and Hoekstra, W. G.,** Selenium, biochemical role as a component of glutathione, *Science*, 179, 588, 1973.
133. **Hadjimarkos, D. M.,** Selenium in man, *N.Y. Med. J.*, 72, 205, 1970.
134. **Frost, D. V.,** Selenium has greater nutritional significance for man; should be cleared for feed, *Feedstuffs*, 44, 58, 1972.
135. **Schroeder, H. A., Frost, D. V., and Balassa, J. J.,** Essential trace elements in man — selenium, *J. Chronic Dis.*, 23, 227, 1970.
136. **Carter, R. F.,** Acute selenium poisoning, *Med. J. Austr.*, 1, 525, 1966.
137. **Shamberger, R. J., Tytko, S., and Willis, R. E.,** Antioxidants and Cancer. II. Selenium and human cancer mortality in the United States, Canada and New Zealand, in *Trace Substances in Environmental Health VIII*, Hemphill, D. D., Ed., Columbia, Mo., 1974, 35.
138. **Schrauzer, G. N., White, D. A., and Schneider, D. J.,** Cancer mortality correlation studies. III. Statistical associations with dietary selenium intakes, *Bioinorg. Chem.*, 7, 23, 1977.

139. **Whanger, P. D.**, Selenium and heavy metal toxicity, in *Selenium in Biology and Medicine*, Spallholz, J. E., Martin, J. L., and Gunther, H. E., Eds., Proc. 2nd Int. Symp. on Selenium, AVI, Westport, Conn., 1981.
140. **Smith, J. C., Zeller, J. A., Brown, E. D., and Ong, S. C.**, Elevated plasma zinc; a heritable anomaly, *Science*, 193, 496, 1976.
141. **Sandstead, H. H.**, Some trace elements which are essential for human nutrition; zinc, copper, manganese and chromium, *Prog. Food Nutr. Sci.*, 1, 371, 1975.
142. **Prasad, A. S.**, Manifestations of zinc abnormalities in human beings, in *Zinc in the Environment, Part 2, Health Effects*, Nriagu, J. O., Ed., John Wiley & Sons, New York, 1980, chap. 2.
143. **Morton, M. S., Elwood, P. C., and Abernathy, M.**, Trace elements in water and congenital malformations of the central nervous system in South Wales, *Br. J. Prev. Soc. Med.*, 30, 36, 1976.
144. **Craun, G. F. and McCabe, L. J.**, Problems associated with metals in drinking water, *J. Am. Water Works Assoc.*, 67, 593, 1975.
145. **Masironi, R.**, Cardiovascular diseases in relation to trace element balance, in *Hardness of Drinking Water and Public Health*, Amavis, R., Hunter, W. J., and Smeets, J. G. P. M., Eds., Pergamon Press, Oxford, 1976, 411.
146. **Stonard, M. D. and Webb, M.**, Influence of dietary cadmium on the distribution of the essential metals copper, zinc and iron on the tissues of the rat, *Chem. Biol. Interact.*, 15, 349, 1976.
147. **Sandstead, H. H.**, Nutrient interactions with toxic elements, in *Toxicology of Trace Elements*, Vol. 2, Goyer, R. A. and Mehlam, M. A., Eds., John Wiley & Sons, New York, 1977, 241.
148. **Demayo, A., Taylor, M. C., and Taylor, K. W.**, Effects of copper on humans, laboratory and farm animals, terrestrial plants and aquatic life, *Crit. Rev. Environ. Control*, 12, 183, 1982.
149. **Berlin, A., Mattielo, G., Allain, P., Ward, M., Sabbioni, E., and Lai, M.**, Aluminum in dialysis fluids and related analytical problems, in *Proc. Int. Conf. Heavy Metals in the Environment*, CEP Consultants, Edinburgh, 1983, 434.
150. **Crapper, D. R., McLachlan, D. R. C., and Deboni, M.**, Aluminum in human brain disease — an overview, *Neurotoxicology*, 1, 3, 1980.
151. **Bowdler, N. C., Beasley, D. G., Fritze, E. C., Goulette, A. M., Hatton, J. D., Ostman, D. L., Rugg, D. J., and Schmittdiel, C. J.**, Behavioral effects of aluminum ingestion on animal and human subjects, *Pharmacol. Biochem. Behav.*, 10, 505, 1979.
152. **Krueger, G. L., Morris, T. K., Suskind, R. R., and Widner, E. M.**, The health effects of aluminum compounds in mammals, *Crit. Rev. Toxicol.*, 13, 1, 1984.
153. **Liu, J. and Stummer, K. L.**, The interaction of aluminum with certain essential metals and its influence upon the pituitary testicular axis, in *Proc. Int. Conf. Heavy Metals in the Environment*, CEP Consultants, Edinburgh, 1983, 442.
154. **Reeves, A. L.**, Species specificity in beryllium carcinogenesis, in *Proc. Int. Conf. Heavy Metals in the Environment*, CEP Consultants, Edinburgh, 1979, 175.
155. **Oehme, F. W.**, Mechanisms of heavy metal inorganic toxicities, in *Toxicity of Heavy Metals in the Environment*, Oehme, F. W., Ed., Marcel Dekker, New York, 1978, chap. 4.
156. **Smith, I. C. and Carson, B. L.**, *Trace Metals in the Environment, Volume 2, Silver*, Ann Arbor Science, Ann Arbor, Mich., 1977.
157. **Schroeder, H. A., Buckman, J., and Balassa, J. J.**, Abnormal trace elements in man: tellurium, *J. Chronic Dis.*, 20, 147, 1967.
158. **Maroni, M., Colombi, A., Buratti, M., Calzaferri, G., Foá, V., Sabbioni, E., and Pietra, R.**, Urinary elimination of vanadium in boiler cleaners, in *Proc. Int. Conf. Heavy Metals in the Environment*, CEP Consultants, Edinburgh, 1983, 66.
159. **Sabboni, E. and Maroni, M.**, A study on vanadium in workers from oil fired power plants, Commission of the European Communities, Luxembourg, 1984.
160. **Phillips, T. D., Nechay, B. R., and Heidelbaugh, N. D.**, Vanadium; chemistry and the kidney, *Fed. Proc. Fed. Am. Soc. Exp. Biol.*, 42, 2969, 1983.
161. **Bengtsson, S. and Tyler, G.**, *Vanadium in the Environment*, Monitoring and Assessment Research Centre, Rep. No. 2., Chelsea College, London, 1976.
162. **Stocks, P.**, On the relations between atmospheric pollution in urban and rural localities and mortality from cancer, bronchitis and pneumonia, with particular reference to 3,4-benzopyrene, beryllium, molybdenum, vanadium and arsenic, *Br. J. Cancer*, 14, 397, 1960.
163. **Hickey, R. J., Schoff, E. P., and Clelland, R. C.**, Relationship between air pollution and certain chronic disease death rates. Multivariate statistical studies, *Arch. Environ. Health*, 15, 728, 1967.
164. **Summers, A. O.**, Bacterial resistance to toxic elements, *Trends Biotechnol.*, 3, 122, 1985.
165. **Trevors, J. T., Oddie, K. M., and Belliveau, B. H.**, Metal resistance in bacteria, *FEMS Microbiol. Rev.*, 32, 39, 1985.
166. **Patrick, F. M. and Loutit, M. W.**, Passage of metals to freshwater fish from their food, *Water Res.*, 12, 395, 1978.

167. **Packham, R.,** Water quality and health, *Water Bull.,* May, p. 6, 1984.
168. **Englande, A. J. and Reimers, R. S.,** Persistence of chemical pollutants in water re-use, in *Water Reuse,* Middlebrooks, E. J., Ed., Ann Arbor Science, Ann Arbor, Mich., 1982.
169. **Davis, R. D.,** Sludge disposal — keeping it safe, *Water Waste Treat.,* Sept., p.38, 1984.
170. **Sherlock, J. C.,** The intake by man of cadmium from sludged land, in *Environmental Effects of Organic and Inorganic Contaminants in Sewage Sludge,* Davis, R. D., Hucker, G., and L'Hermite, P., Eds., D. Reidel, Dordrecht, 1983, 113.
171. **Davis, R. D., Carlton-Smith, C. H., Johnson, D., and Stark, J. H.,** Evaluation of the effects of metals in sewage sludge disposal, *Water Pollut. Control.,* 84, 380, 1985.
172. **Baxter, J. C., Barry, B., Johnson, D. E., and Keinholz, E. W.,** Heavy metal retention in cattle tissues from ingestion of sewage sludge, *J. Environ. Qual.,* 11, 616, 1982.
173. **Bryan, G. W.,** The occurrence and seasonal variation of trace metals in the scallops *Pecten maximus* (L.) and *Chlamys opercularis* (L.), *J. Mar. Biol. Assoc. U.K.,* 53, 145, 1973.
174. **Southworth, G. R., Parkhurst, B. R., Herbes, S. T., and Tsai, S-C.,** The risk of chemicals to aquatic environments, in *Environmental Risk Analysis for Chemicals,* Conway, R. A., Ed., Van Nostrand Rheinhold, New York, 1982, chap. 4.

Chapter 2

SOURCES OF HEAVY METALS IN WASTEWATER

T. Stephenson

TABLE OF CONTENTS

I.	Introduction		32
II.	Sources of Heavy Metals		33
	A.	Anthropogenic and Nonanthropogenic Sources	33
		1. Total Environmental Inputs	33
		2. Occurrence in the Lithosphere	34
		3. Mine Production	34
	B.	Geochemical Background	37
	C.	Atmospheric Deposition	38
	D.	Runoff	40
		1. Road Runoff	41
		a. Vehicle Lubricant Loss	42
		b. Vehicle Tire Degradation	42
		c. Vehicle Exhaust Emission	42
		d. Cleaning and Deicing	43
		2. Storm Events	43
	E.	Domestic Discharges	44
	F.	Industrial Discharges	45
		1. Metal Uses	45
		a. Aluminum	45
		b. Antimony	47
		c. Arsenic	47
		d. Beryllium	47
		e. Bismuth	47
		f. Cadmium	47
		g. Chromium	47
		h. Cobalt	48
		i. Copper	48
		j. Lead	48
		k. Manganese	48
		l. Mercury	48
		m. Molybdenum	49
		n. Nickel	49
		o. Selenium	49
		p. Silver	49
		q. Tellurium	49
		r. Thallium	49
		s. Tin	49
		t. Vanadium	49
		u. Zinc	49
		2. Selected Industries	50
		a. Inorganic Chemicals Manufacture	50

	b.	Electroplating..50
	c.	Metal Finishing...52
	d.	Photoprocessing 52
	e.	Paints, Inks, and Pigments 53

III. Heavy Metals in Wastewaters......................................53
    A. Source Contributions 53
    B. Case Studies...55
        1. Camden County, N.J...................................55
        2. Kokomo, Ind..57

References..59

I. INTRODUCTION

Heavy metals that occur in wastewater-treatment processes will have originated from Earth's crust in the first instance. The quantity of any one particular metal present in raw sewage influent to a treatment works is dependent on a combination of many factors. These can include, for example, the geographical location of the wastewater-treatment plant, the type of sewerage system served, and the presence or absence of domestic and industrial discharges. However, before such local conditions affect metal concentrations, there are primary factors that will influence the appearance of any metal in sewage and these include chemical properties, abundance, availability of ores, and usefulness to man. The chemistry of a metal, discussed in Chapter 1, will determine to some extent the type of ores present and its ultimate use by man. Abundance in the lithosphere (Earth's crust) may affect metal appearance in the hydrosphere (water environment) and also quantities of extractable ores. The availability of such metal ores will, in turn, have some influence on the quantities that are used by man.

Sources of metals that enter the water environment can be divided into two main categories: those that originate from man's activities and those of natural origin. The term applied to the former category is *anthropogenic,* although the strict meaning of anthropogeny is the study of the origin of man, rather than the study of things originating from man. It has been suggested that *anthropurgic* would be a more correct term etymologically,[1] yet anthropogenic is in widespread modern usage and has recently been defined as "referring to environmental alterations resulting from the presence or activity of humans."[2] Therefore metals from natural sources can be defined as nonanthropogenic.

Sources of metals that can eventually reach wastewater-treatment works can be considered as five types: domestic and industrial effluents, runoff, atmosphere, and lithosphere. Domestic and industrial discharges are probably the two most important anthropogenic sources, and anthropogenic metals may also come from the atmosphere and runoff, depending on whether the sewerage system is separate or combined with storm drainage. Nonanthropogenic sources can also include the atmosphere and runoff, the latter arising from geochemical background concentrations due to sources in the lithosphere. The interrelationship of the major sources of heavy metals and the possible pathways to wastewater-treatment processes are shown in Figure 1.

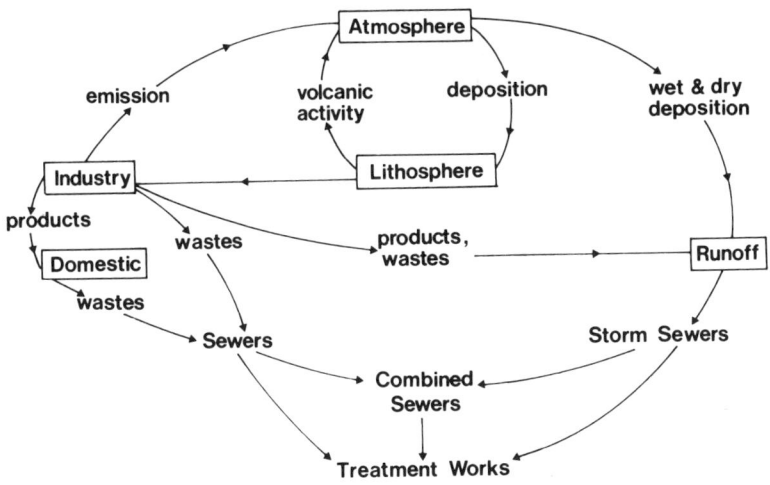

FIGURE 1. Sources and pathways of heavy metals entering wastewater-treatment processes.

II. SOURCES OF HEAVY METALS

A. Anthropogenic and Nonanthropogenic Sources
1. Total Environmental Inputs

Over the last century there has been a sharp increase in environmental contamination by many heavy metals; e.g., lead[3] and mercury.[4] In general, background concentrations of metals such as antimony, cadmium, chromium, copper, lead, mercury, tin, thallium, and zinc have probably been raised globally as a result of human activities, leading to increased concentrations existing in the atmosphere[5,6] and hydrosphere.[7,8] Thus several authors have estimated the total input of a particular metal to the environment, including anthropogenic and nonanthropogenic sources.

It has been stated that between 80,000 and 250,000 t/year of copper reaches the world's environment due to natural weathering processes.[9] This nonanthropogenic source is between 40 and 67% of the total copper entering the global environment.[9] A breakdown of the anthropogenic sources has indicated that sewage makes a major contribution, followed by the processing of copper ore — most of which is in mine tailings. The estimated natural annual input of zinc to the environment is greater than that of copper, accounting for 720,000 t by weathering processes.[10] Erosion and other natural sources such as sea-salt sprays, forest fires, vegetation, and volcanoes contribute a further 43,500 t/year. The contribution of nonanthropogenic sources to the total global environmental load of zinc is approximately 65%, at the upper end of the range estimated for copper. Total anthropogenic inputs of zinc have been calculated as 414,000 t/year, of which approximately half can be accounted for in sewage and primary zinc production. Weathering phenomena can also account for a significant proportion of lead released to the global environment and this source has been estimated at between 21,000 and 180,000 t/year.[11] However, it should be noted that anthropogenic emmissions to the atmosphere from antiknock agents present in internal combustion engine fuel can contribute 333,000 t/year.[11]

Hutton[12] summarized the sources of cadmium inputs to the environment in European Economic Community countries (Table 1). Volcanic action was the only natural source, with a contribution amounting to only 15% of the inputs to the atmosphere. Atmospheric cadmium inputs were estimated to be only approximately 4% of the total, with land and water receiving 88 and 8%, respectively. Anthropogenic inputs of cadmium to the environment of Switzerland total approximately 30 t/year out of 130 t/year brought into the country.[13]

Table 1
INPUTS OF CADMIUM TO THE EUROPEAN COMMUNITY ENVIRONMENT

	Inputs (t/year)		
Source	Air	Land	Water
Volcanic action	20	ND	ND
Nonferrous metal production			
Zinc and cadmium	20	200	50
Copper	6	15	ND
Lead	7	40	20
Production of cadmium-containing materials	3	90	108
Iron and steel production	34	349	ND
Fuel combustion			
Coal and lignite	8	390	ND
Oil and gas	0.5	14.5	—
Waste disposal	31	1434	ND
Sewage sludge disposal	2	130	62
Phosphate fertilizers	—	346	62
Totals	**132**	**3009**	**273**

Note: ND, not determined.

From Hutton, M., *Ecotoxicol. Environ. Saf.*, 7, 9, 1983. With permission.

Of this, 6 t was lost to the atmosphere by incineration operations, 4 t from the metal industry, and 3 t was disposed of in sewage sludge treatment.

2. Occurrence in the Lithosphere

The abundance of any element within Earth's crust or lithosphere will be a factor in determining its use by man and is usually quoted on a percentage by weight basis. Oxygen is the most abundant element in the lithosphere, and its concentration has been estimated at between 464,000 and 495,000 ppm.[14] Thus oxygen is far more ubiquitous than the majority of heavy metals, of which only aluminum, iron, manganese, and vanadium with abundances of 82,300, 56,300, 950, and 135 ppm, respectively, appear in the top 20 most abundant elements.[15] Indeed, aluminum and iron are the third and fourth most common elements in the lithosphere after oxygen and silicon. Heavy metals with abundances >10 ppm also include chromium, nickel, zinc, copper, cobalt, and lead with abundances of 100, 75, 70, 55, 25, and 12.5 ppm, respectively. Beryllium, tin, and molybdenum have abundances >1 ppm as well, followed by the rest of the heavy metals with concentrations <1 ppm. Therefore, the order of abundances of heavy metals in the lithosphere according to Taylor[15] is

$$Al > Fe > Mn > V > Cr > Ni > Zn > Cu > Co > Pb > Be > Sn > Mo > Tl > Sb, \quad Cd > Bi > Hg > Ag > Se > Te$$

The heavy metals are extracted from ores in which the percentage of metal on a weight for weight basis will be far greater than the average crustal abundance.

3. Mine Production

Almost all anthropogenic inputs of heavy metals that have environmental significance

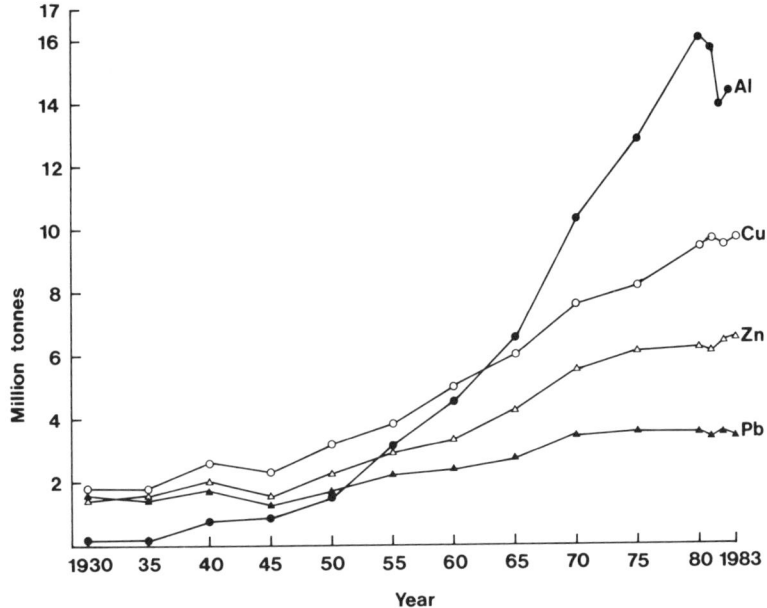

FIGURE 2. Total global mine production of aluminium (●), copper (○), lead (▲), and zinc (△) from 1930 to 1983.[18]

ultimately come from deliberate mining operations. Production of metals by man from ores in Earth's crust has increased with the development of civilization. Settle and Patterson[16] charted the rise in the production of lead over the last 5,000 years. Upon the discovery of cupellation in about 3000 B.C., lead production increased to above 10 t/year and was >100 t/year around 2,500 years later when lead coinage was in use. When the Roman lead mines were exhausted after 0 A.D., production was two orders of magnitude greater at 10,000 t/year. Today lead production is more than 2,000,000 t/year, with mine production reaching a peak of 2,544,000 t/year in 1979.[17]

Production from mining, or that from smelting and refining processes, for many economically important metals has increased dramatically over the last 50 years. Iron production has for many years been in excess of any other metal, rising from 182,000,000 t/year in 1930 to a peak of 912,000,000 t/year in 1979, a fivefold increase.[18] Currently aluminum is produced in quantities second only to iron, with more than 16,000,000 t/year in 1979 (Figure 2). However, in 1930 smelter production of aluminum was a mere 219 t/year and was much less than that of copper, lead, and zinc (Figure 2). Over the last 50 years, modern technology has found increasing uses for aluminum, and more efficient mining and refining techniques have also encouraged its widespread utilization. In 1930 the annual amounts of copper, lead, and zinc produced were approximately equal and almost an order of magnitude greater than aluminum. Up until 1980, the yearly productions of copper and zinc have increased by approximately five- and fourfold, respectively, whereas during the same period, lead annual mine production has only doubled. The figures shown in Figure 2 demonstrate that man's relative demands for different metals have changed rapidly over the last 50 years, thus affecting the total environmental input.

Cadmium and cobalt productions have expanded since 1935 and up until 1980 had increased by approximately 10 and 16 times, respectively, as shown in Figure 3. In contrast the annual production figures for mercury and silver, also shown in Figure 3, had only increased by 2 and $1\frac{1}{2}$ times, respectively, over the same period. Other metals, including beryllium, tellurium, and tin have also not been mined to a much larger degree over the

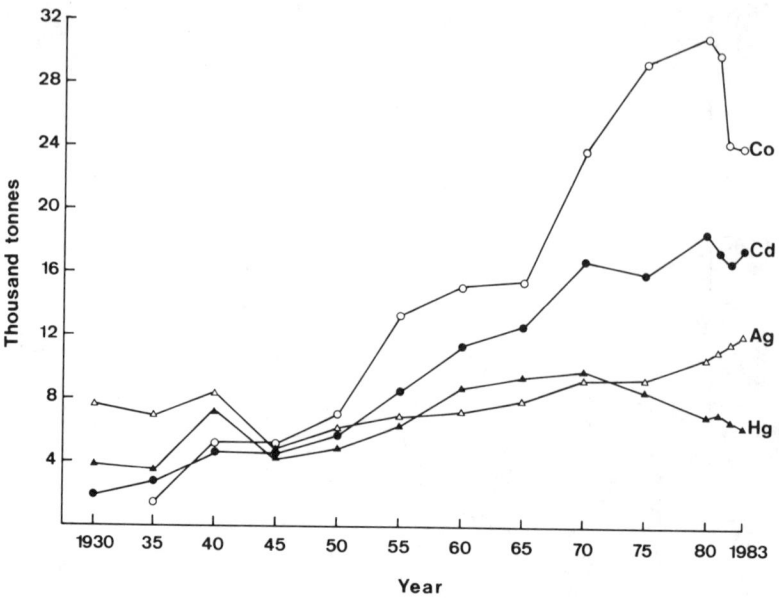

FIGURE 3. Total global mine production of cadmium (●), cobalt (○), mercury (▲), and silver (△) from 1930 to 1983.[18]

past 50 years. Beryllium production only increased by $2^1/_2$ times between 1935 and 1983, tin by a mere 21% between 1935 and 1983, and tellurium also by $2^1/_2$ times between 1940 and 1983. Production of bismuth has expanded from 1300 t/year in 1938 to 4055 t/year in 1983, a threefold rise.[18] Antimony mine production has become greater by almost five times in 50 years; it was 12,900 t in 1930 and 63,500 t in 1980.[18] Nickel and selenium are two other metals for which production has increased dramatically over the same half century. The tonnage of mined nickel expanded by more than 12 times and that of selenium by almost 6 times. Between 1945 and 1980, the total global production of manganese rose from 386,100 t/year to 29,000,000 t/year; i.e., almost 8 times. Mined chromite ore went from 559,000 t/year in 1930 to 11,237,000 t/year in 1980, an increase of some 20-fold.

Therefore almost all metals reached peak production around 1980, and the expansion in mined quantities was especially spectacular for metals such as aluminum, molybdenum, and vanadium. At the opposite end of the spectrum, tin, silver, and mercury production rose only slowly. The order of increase in mine production between 1930 and 1980 is

Al > Mo > V > Co > Cr > Ni > Cd > Mn > Se > Te > Cu > Fe > Sb > Zn > Bi > Be > Pb > Hg > Ag > Sn

Thus the potential anthropogenic inputs to the environment, and therefore wastewater treatment, have in general terms increased at a greater rate for those metals near the top of the list. However, this ranking does not indicate the magnitude of the potential anthropogenic discharge of each metal. The order in terms of tonnage of metal mined, smelted, or refined in 1980 is

Fe > Mn > Al > Cu > Zn > Pb > Cr > Ni > Sn > Mo > Sb > V > Co > Cd > Ag > Hg > Bi > Se > Te > Be

This is probably more important in determining the significance of heavy metals in wastewater

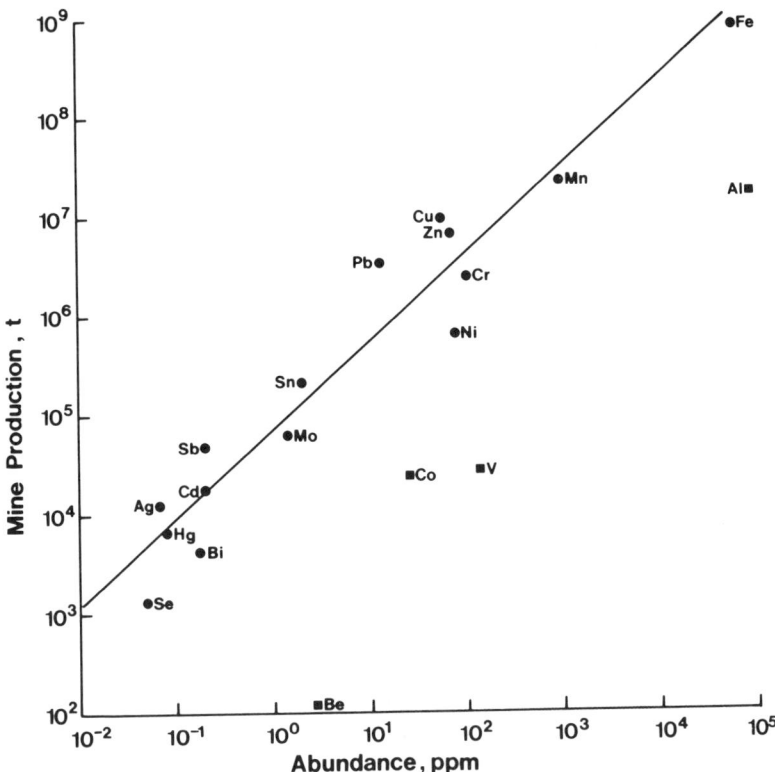

FIGURE 4. Log-log plot of heavy-metal abundance in the lithosphere[15] against total global mine production in 1983.[18]

treatment. Indeed, the majority of studies of metal behavior in sewage treatment systems concentrate on six metals[19] that include chromium, copper, lead, nickel, and zinc which are ranked from fourth to eighth in terms of mine production. Nevertheless, the rate of increase in mine production of metals such as cobalt, molybdenum, and vanadium indicate that they could become of increasing environmental concern and significance in wastewater treatment.

The crustal abundance[15] of the heavy metals can be compared to their total mined productions.[18] A plot of 19 metals indicates that there is a log-log relationship between the two parameters using production quantities for 1983 for the 15 metals antimony, bismuth, cadmium, chromium, copper, iron, lead, manganese, mercury, molybdenum, nickel, selenium, silver, tin, and zinc (Figure 4). The regression coefficient (r) for this line was very good, $r = 0.97$. This demonstrates that, in general, it is the overall crustal abundance of any metal that will in some way ultimately determine the amount that is extracted by man. It should be noted that aluminum, beryllium, cobalt, and vanadium were well below the regression line, possibly indicating that these metals have not yet reached their maximum output. In addition, the production of several metals above the line but included in the regression, including lead, silver, and tin, has not increased dramatically over the preceding 50 years.

B. Geochemical Background

The natural composition of soils varies greatly with geographical location and as a result so can the heavy-metal content. This in turn can effect the nonanthropogenic metal concentration in the hydrosphere. Weathering and erosion can account for as much as 750,000 t/year of zinc inputs to the environment.[6] The majority of this zinc comes from shale rock which has a high weathering rate[20] of 6,500,000 t/year and a high zinc content of 95 mg/

Table 2
CONCENTRATIONS OF HEAVY METALS IN WET DEPOSITION FROM REMOTE SITES

Location	Metal concentration (ng/ℓ)									Ref.
	Sb	As	Cd	Cu	Pb	Mn	Hg	Ag	V	Zn	
Greenland	34	19	639	850	—	250	48	—	—	1100	26
Greenland	—	—	8	—	140	177	—	—	16	219	27
Antarctica	—	—	7	47	40	18	—	6	—	7	28

kg, compared to 70, 20, and 16 mg/kg for igneous, limestone, and sandstone, respectively.[21] This is also true for lead inputs.[11] Concentrations of selenium in soil can be so high as to pose a health threat — in some parts of North America consumption of plants growing on such soils has proved fatal to livestock.[22] Other areas of the world, especially New Zealand, are particularly deficient in selenium.

High geological background concentrations of metals can affect the levels present in bodies of water but are unlikely to be of great importance as a source of heavy metals in wastewater. The main pathway of entry of such metals into sewerage systems is due to groundwater infiltration. In general, rates of infiltration may range[23] from 0.2 to 28 m^3/ha/day, depending on the quality of the sewers and their elevation compared to the groundwater. In urban areas, the impervious nature of much of the surface results in decreased infiltration.

C. Atmospheric Deposition

Atmospheric deposition of heavy metals is made up from two fractions: dry fallout and wet deposition. Dry fallout has been defined as matter that can be deposited into an open sample container in the absence of any hail, snow, or rain.[24] Wet deposition covers all forms of precipitation, which includes hail, snow, and rain. The sum of these two fractions is usually referred to as bulk deposition. Determination of metals in deposition has indicated that the contribution from the dry fraction can be of significance. Dry deposition has been reported as accounting for between 30% for lead and 50% for manganese and nickel of bulk deposition at rural sites with an even greater range of between 20% for arsenic and lead and 60% for cadmium of bulk deposition at urban sites.[25]

Heavy metals in the atmosphere originate from both anthropogenic and nonanthropogenic sources, as previously discussed. Measurement of trace metals in the atmosphere and in deposition at sites far and close to areas of inhabitation has unequivocally demonstrated the influence of man on metal concentrations. Galloway et al.[25] surveyed the literature on trace metals in atmospheric deposition and divided the results among three major categories: remote, rural, and urban. Remote sites included areas of lowest concentration, which basically meant the Arctic and the Antarctic. Rural sites were defined as those that demonstrated a baseline concentration for a region without the immediate influence of anthropogenically produced atmospheric metals. Urban sites covered any location, whether in a town or away from it, which were immediately influenced by emissions of an anthropogenic nature.

Studies of metal concentrations in snow and ice from remote areas of the globe demonstrate that background concentrations of metals are often in the nanograms-per-liter range, as shown in Table 2. However, it is evident that the levels of metals detected from the Greenland sites[26,27] were higher than those from the Antarctic,[28] probably due to the proximity of the former to the major industrialized nations of Europe and North America.[28] Copper and lead concentrations in the southern polar regions were greater than those of metals such as manganese and zinc (Table 2), again possibly due to the high anthropogenic emissions of

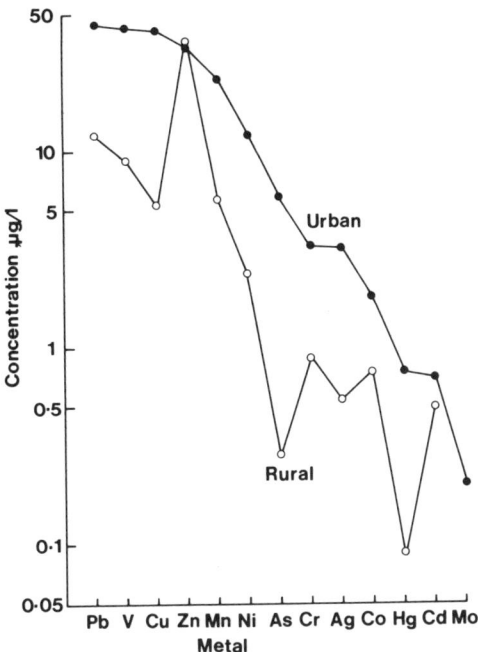

FIGURE 5. Median concentrations of heavy metals in wet deposition from results presented in the literature.[25]

these metals. Galloway et al.[25] estimated that the median concentrations of heavy metals in precipitation over remote sites were, in general, approximately an order of magnitude less than in rural wet deposition.

The elemental concentration in air near ground level at Wraymires, a rural site in the Lake District, U.K., was measured for a total of nine years between 1971 and 1979.[29] Over this period the concentrations of cobalt, lead, and zinc decreased by approximately 10 to 20% annually. Other metals, including aluminum, manganese, and vanadium have also decreased over the same period. Indeed, for a total of 36 elements (including nonmetals) there was no significant increases in atmospheric levels; this was also true for 3 other sites studied between 1972 and 1979.[29] Concentrations of cadmium and lead at four forested sites in the southern U.S. also decreased from 1976/1977 to 1980/1982.[30] Such decreases in atmospheric metal concentrations in deposition from rural sites are probably due to a reduction in metal emissions during the periods of study, especially as analysis of metals in atmospheric particulates at rural sites has indicated that they were anthropogenic in origin.[31]

Metal concentrations in wet deposition in urban areas are approximately two to ten times greater than at rural sites (Figure 5) and in certain cases even higher; e.g., close to power stations using fossil fuels.[32] A study of particulates in the atmosphere in West Germany demonstrated that the greatest amounts of beryllium, chromium, cobalt, and nickel were found in two urban areas compared to rural holiday resorts.[33] Chromium, cobalt, and nickel concentrations were approximately 14, 5, and 42 ng/m^3, respectively, in a city street canyon — about twice as much as detected at a residential site in the same city. Roadside concentrations of cadmium in a town center were two times greater than at a rural location in northwest England.[34] Atmospheric mercury concentrations at rural sites in Tuscany, Italy were 4.0 ng/m^3, compared to 8.3 to 16.1 ng/m^3 measured in three Italian cities.[35] However, close to a chlor-alkali plant, the mean mercury levels ranged from 22.1 to 73.2 ng/m^3. Therefore the most important factor in determining the level of atmospheric heavy-metal pollution is the proximity to a large-scale industrial process without proper emission controls.

Nürnberg et al.[36] monitored cadmium, copper, lead, and zinc at 20 locations in West Germany and found that concentrations were greatest in Goslar and Stolberg, sites associated with cadmium and zinc production and large lead smelters. At Goslar in 1981, the average daily wet depositions for cadmium, copper, lead, and zinc were 9, 30, 152, and 1535 $\mu g/m^2/$day, respectively, compared to 2.3, 27, 158, and 300 $\mu g/m^2/$day, respectively for Essen, the most heavily polluted general urban area. The areas close to power stations will also have comparatively high concentrations of metals, depending on the type of fossil fuel used. Fly ash originating from coal combustion contains considerable quantities of beryllium,[37] whereas oil contains high concentrations of vanadium.[38] Dewling et al.[39] demonstrated that the incineration of sewage sludge could have a considerable impact upon local atmospheric air pollution by heavy metals. However, it is possible to neutralize such local pollutant sources with the application of proper emission-control technology. In one case study, analysis of soil in the vicinity of a lead smelter in northern England revealed a history of lead pollution.[40] At the then-current rate of atmospheric lead deposition, it would have taken 72 years for the soil lead concentrations to become that high, yet the site had only been operational for 18 years. Therefore it was concluded that past uncontrolled lead emissions accounted for the soil pollution and that stack filters and other source controls had limited atmospheric pollution since. However, lead deposition was still relatively high at 374 mg/m^2/year.

Direct input of heavy metals from atmospheric sources into wastewater is unimportant since sewers are covered and the surface area exposed at treatment works is relatively small. A conventional wastewater-treatment works serving a population of approximately 200,000 (such as the Hogsmill Valley Works, Thames Water Authority in the U.K.,[41]) would have a surface area of approximately 8000 m^2; i.e., 0.8 ha. Assuming that the plant was sited in an area of high atmospheric pollution, the annual bulk deposition rate of lead, for example, could be from 2.11 to 2.15 kg/ha/hr.[42,43] Therefore a maximum of 1.72 kg/year of lead would be deposited directly into the sewage at the works. A treatment works of this size would receive approximately 59,000 m^3 of raw sewage every day; i.e., 21,500,000 m^3/year. A typical raw sewage lead concentration would be 0.395 mg/ℓ,[44] so that the annual input to the works would be 8490 kg. Therefore it is obvious that even at a high rate of atmospheric lead deposition, only in the region of 0.02% of the total metal loading would be due to direct atmospheric input. The total area served by a works such as Hogsmill Valley would be about 72 km^2 (7200 ha), of which approximately half would be urbanized. Only about 7% of the built-up areas would be street surface and other impervious coverings,[45] making a total area discharging to the treatment works of 252 ha. If the area as a whole suffered from quite high air pollution, lead deposition would be 0.35 kg/ha/year as in Walsall,[46] U.K. Therefore, if 100% of metal deposited on impervious surfaces were washed into combined storm sewers and eventually reached the wastewater-treatment plant, a total of 88 kg of lead from atmospheric sources would be present in the raw sewage; i.e., approximately 1% of the total lead concentration. It is apparent that over a long period, the contribution of atmospheric metals to the total metal loading at most wastewater-treatment works is probably negligible. Over the short term, however, the indirect contribution may be more significant due to runoff from storms.[47]

D. Runoff

Land use patterns will be a major factor in determining the concentration of heavy metals in runoff. Sonzogni et al.[48] estimated that the highest concentrations of copper, lead, and zinc in runoff originated from industrial sites, and lower concentrations were in general associated with rural runoff (Table 3). Water that originated from the surfaces of roads and highways constitutes a major source of pollution in urban runoff. Road surfaces in built-up areas present a large area of relatively impervious material that is exposed to a variety of

Table 3
RANGES OF UNIT AREA LOADINGS OF HEAVY METALS IN RUNOFF BY LAND USE

	Metal loadings (g/ha/year)					
	Cu		Pb		Zn	
Land use	Min.	Max.	Min.	Max.	Min.	Max.
Rural						
Mixed agriculture	2	900	2	80	5	300
Arable	14	64	5	6	26	83
Improved pasture	21	38	4	15	19	172
Forest	20	30	10	30	10	30
Wasteland	20	30	10	30	10	30
Urban						
General	2	210	140	500	300	1000
Restricted	30	30	60	60	20	20
Commercial	70	130	170	1100	250	430
Industrial	290	1300	2200	7000	3500	12000

Reprinted in part with permission from *Environ. Sci. Technol.*, 14(2), 149. Copyright 1980, American Chemical Society.

pollutant sources. As these areas are normally fairly free draining, the relative contribution of road-surface runoff to total urban runoff is often high. A study in Oklahoma City, Okla. indicated that the volumetric flow of urban runoff was approximately 12.5% of total precipitation,[45] with the rest presumably lost due to evaporation and infiltration. Only 7.16% of the city area was street surface, yet the runoff from this source was responsible for 57% of the total runoff. In addition to street surfaces being scoured by rainfall and snowmelt, street-cleaning activities may also cause surface pollutants to be washed into the runoff.

1. Road Runoff

Sources of road-surface pollutants can, according to Pope et al.,[49] be divided into seven main categories. Each of these categories will contribute varying amounts of heavy metals to the runoff, depending on the particular metal. The seven categories are vehicle lubricant loss; load loss; tire degradation; exhaust emission; road-surface degradation; road cleaning and deicing, and atmospheric precipitation. Some of the more important sources are discussed in further detail below. It should be remembered that the problem of atmospheric lead pollution from tetraethyl lead added to gasoline as an "anti-knock" agent to prevent preignition was recognized as long as 50 years ago.[50,51] In addition, the toxic effects of lead and its possible harmful effects on human beings, especially in soft water and highly urbanized areas, have been the subject of considerable attention.[52-54] It is probably for these two reasons that the majority of studies of heavy metals in road runoff have concentrated on lead, as summarized in the review by Laxen and Harrison.[55]

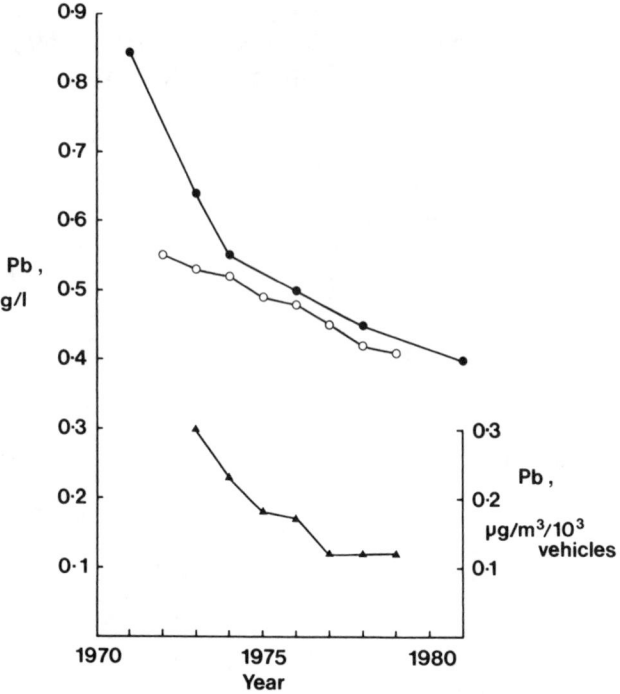

FIGURE 6. Maximum permitted lead content (●) and actual lead content (○) of petrol in the U.K. compared to atmospheric lead concentrations (▲) along an urban motorway.[60]

a. Vehicle Lubricant Loss

During the driving cycle, not all the tetraethyl lead will be exhausted, and some will be retained in the engine oil.[55] Due to lubricant leaks from head gaskets and sumps, some of this lead is deposited onto the road surface.[56] This is in addition to any metals, such as vanadium, that are already present in the oil formulation.

b. Vehicle Tire Degradation

In one study of an elevated section of motorway, the mass of particulate rubber lost to the road surface was estimated as being 0.14 kg/m^2 from quantities collected in the runoff.[57] This figure was in good agreement with the 0.12 kg/m^2 reached by using the total vehicular mileage and tire wear, assuming mean replacement to be 46,000 km per tire.[57] It has been estimated that the total mass of rubber released to the environment in the U.S. is >270,000 t/year.[58] The zinc content of vehicle tire tread is approximately 1% on a weight-to-weight basis[57] and therefore >2,700 t/year is lost to the environment in the U.S. through tire wear alone, with the majority of this directly onto the road surface and therefore into the runoff.

c. Vehicle Exhaust Emission

The release of heavy metals, in particular lead, to the atmosphere via vehicle exhaust is one of the most important sources of heavy metals in road runoff.[59] The lead content of gasoline is now restricted by law in most countries due to the increasing awareness of the possible harmful health effects of this metal. Since the beginning of the 1970s, the maximum permissible concentration of lead in fuel in the U.K. has been steadily decreased and this has resulted in a decrease in the actual concentration (Figure 6).[60] As a consequence, the atmospheric concentration of lead near major roads has actually decreased, as also shown in Figure 6. Studies on the atmospheric concentration of lead have demonstrated clearly the link between vehicle usage and occurrence of the metal. Measurements of particulate lead in the atmosphere on the M4 Motorway at Harlington, London, have shown a good correlation between traffic flow and lead concentrations.[61] At night, when traffic flows are <500 vehicles per hour, the lead concentration decreases to <2 µg/m^3; yet when flow is >5000 vehicles per hour, the lead concentration can be almost 25 µg/m^3. Such atmospheric lead concentrations occur during dry weather. During periods of atmospheric precipitation, this lead will be "washed" from the atmosphere onto the surface of the road, thus causing a marked decrease in the atmospheric lead concentration.[61] In this way atmospheric lead from vehicle-exhaust gases contributes towards the lead content of road runoff.

d. Cleaning and Deicing

The major materials deliberately added to road surfaces are grit and salt for deicing purposes. However, little or no grit is normally added to major roads in the U.K. because it can be a major hazard, causing windscreen breakages in fast and heavy traffic conditions.[57] In addition, such admixture grit is often uneconomic (due to the subsequent drainage system cleaning) and therefore pure rock salt is often used. Rock salt contains significant quantities of heavy-metal contaminants, including approximately 12 mg/kg nickel and 9 mg/kg chromium, as well as considerable quantities of iron.[57] In a study of runoff from a motorway in central England, Hedley and Lockley[57] found that the major part of the annual loading of chromium and nickel was derived from the road-salting operations.

2. Storm Events

Concentrations of heavy metals in runoff during storm events have been studied at several sites (Table 4). Higher concentrations of metals in storm runoff have been noted from road sites[49,57,59] compared to urban areas.[63-65] This is probably due to the fact that the whole area drained from a road surface is impervious, whereas a lesser proportion of urban areas are impervious. Nearly all of the metal present in such runoff is derived from that associated with street dusts which contain elevated concentrations of heavy metals.[65,68] Indeed, Colwill et al.[59] studied the association of lead with different particle sizes in road dust and in runoff sediment and found that the patterns of particulate sizes were similar (Figure 7). It was observed that high lead concentrations were associated with smaller particle sizes, which is also true for copper.[69] Harrison and Wilson[69] also found that there were significant correlations (at the 0.1 significance level) between suspended solids concentrations and the metals copper and lead. This is further evidence that lead on road surfaces is derived from the atmosphere, as >90% of lead in the air at rural, urban, and motorway sites has been found to be particulate.[70] It is possible, however, that metals associated with the solid fraction of road dust could be leached off by rainwater, with cadmium more extractable than manganese, which was in turn more extractable than lead.[71] Indeed cadmium, as well as copper, was associated with colloidal materials on the whole.[72] It is apparent that the mobilization of metals from the surfaces of roads is dependent on their physicochemical speciation.[72] Soluble heavy metals are removed early in a storm event whereas particulate-associated metals require a strong flow to remove them into drains. In addition changes in pH during the runoff event to affect the species distribution of some metals, in particular lead.

Roberts et al.[47] observed the changes in the concentrations of heavy metals in the influent at a wastewater-treatment works receiving storm water runoff. During one storm event, the volumetric flow increased from approximately 1.8 × dry weather flow (dwf) to almost 4 dwf in 1 hr before falling to <1.5 dwf after a further 2 hr. The relative metal loadings as multiples of the dry-weather loadings increased by larger fractions than the flow. Zinc loading peaked at >5 × the dry-weather loading, the maximum copper loading occurred at the same time as the highest flow and was >8 × the dry-weather loading, and the maximum lead loading was >30 × that during dry weather. These large increases in heavy-metal

Table 4
CONCENTRATIONS OF HEAVY METALS IN URBAN AND ROAD RUNOFF DURING STORM EVENTS

Location	Metal (µg/ℓ)						Ref.
	Cd	Cr	Cu	Pb	Ni	Zn	
Urban, Zürich	0.8	3	14	140	—	160	63
Urban, Manassas, Va.	—	—	—	750	—	—	64
Urban, London	min. 4.1 max. 11.4	— —	75 118	130 386	— —	186 287	65[a]
Urban, Finland	min.- max.-	— —	10 680	13 1500	— —	39 1700	66[b]
Road, M6, U.K.	min.- max.-	— —	100 1440	450 4940	— —	650 8010	57[c]
Road, M1, U.K.	min. 1.6 max. 4.9	55 215	90 193	529 1576	84 232	23 422	49[d]
Road, M1, U.K.	min. <3.0 max. 102.0	18 85	7 30	100 8015	— —	116 4047	59
Road, M6, U.K.	<0.45 µm 1.4 >0.45 µm 1.1	— —	27 48	18 250	— —	— —	67[e]

[a] Two storm events.
[b] Seven sites.
[c] Monthly means in 1 year.
[d] Four times in a single storm event, insoluble fraction only.
[e] Median concentrations, ten storm events.

concentrations were the result of metals being flushed from the road surfaces during the early part of the storm and demonstrate that, in areas of combined sewerage systems, storm runoff can have a significant impact on metals received at wastewater treatment works.

E. Domestic Discharges

The heavy metals present in domestic wastewater will be derived from all discharges from residential areas and includes service industries such as automobile repair shops and restaurants. However, purely household wastewater can contain a significant quantity of metals. Zinc is the most abundant metal present in human faeces and occurs at concentrations of approximately 250 mg/kg dry solids.[73] Fecal concentrations of copper are approximately 68 mg/kg followed by lead, nickel, and cadmium at 11, 4.7, and 2 mg/kg, respectively.[73,74] These concentrations have been compared to the heavy-metal of nonindustrial sewage sludge from a survey of 17 small domestic works in the east of England.[75] The median concentrations of zinc and cadmium were just over twice that in feces; whereas the copper, nickel, and lead contents were five to ten times greater. These differences illustrate that there are many other sources of heavy metals in domestic sewage other than human wastes.

It is a fact that mundane daily domestic tasks, such as taking a bath, brushing teeth, and washing hair, contribute a significant proportion of metals discharged to sewers in residential

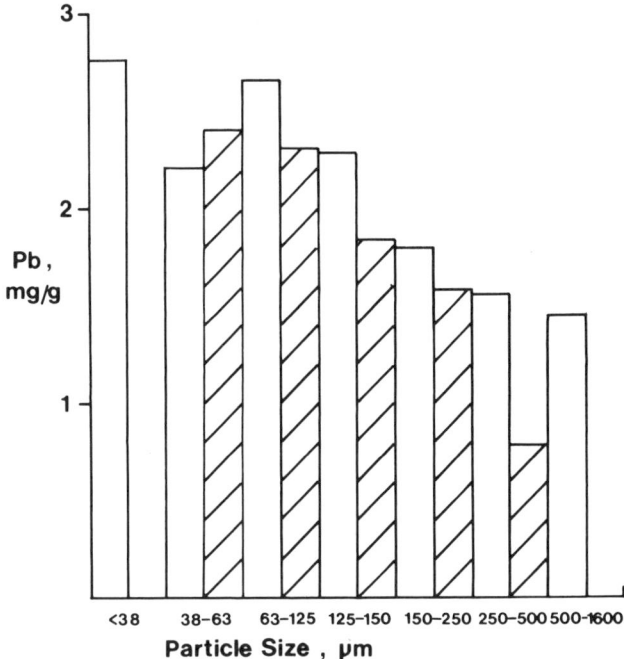

FIGURE 7. Lead content of road dust and runoff sediments by particle size.[59]

areas.[76] Household products such as cleaners, cosmetics, medicines, and polishes can all contain heavy metals. Deodorants that contain aluminum compounds and cosmetics that have a wide variety of metals in their formulations are all likely to be washed into the sewerage system during bathing. Tin(II) fluoride is now a popular ingredient of toothpaste, and it has been estimated that approximately 1,000,000 kg of tin from this source alone is released in the U.S. in a year.[58] Some antidandruff shampoos can contain up to 1% selenium sulfide.[77] Atkins and Hawley[76] categorized sources of pollutants in domestic sewage into 29 different types, of which 22 contain heavy metals at some sort of concentration (Table 5). It should be noted that this inventory of products includes those used in hotels and restaurants, hospitals, laboratories, car washes and garages, warehouses, and shopping centers that would not normally be considered as industrial discharges. Heavy metals such as aluminum, copper, iron, lead, and zinc appear in a wide variety of products (Table 5) and therefore it is to be expected that these metals appear in relatively high concentrations in domestic wastewater. Aluminum is used in toothpastes, antiperspirants and deodorants, eye shadow, astringents, and barrier creams in the cosmetics category alone — all products eventually being disposed of down the drain. Zinc is also used in many cosmetic products, as well as in many pesticides, paints, pigments, and polishes.

F. Industrial Discharges
1. Metal Uses

Heavy metals have a huge variety of industrial applications which will to some extent influence their appearance in wastewaters. The uses of metals have previously been briefly summarized by Patterson[78] and Matthews[79] and each is considered in detail below.

a. Aluminum

The metal is lightweight and is used in the construction of buildings, in transport, and in

Table 5
SOURCES OF HEAVY METALS IN DOMESTIC WASTEWATER BY PRODUCT TYPE[76]

	Al	Sb	As	Be	Bi	Cd	Cr	Co	Cu	Fe	Pb	Mn	Hg	Mo	Ni	Se	Ag	Sn	Ti	V	Zn
Automotive products	X		X				X	X		X	X			X					X		X
Caulking compounds	X		X	X			X	X		X	X								X		X
Cleaners	X						X		X	X									X		X
Cosmetics	X			X	X	X		X	X	X		X			X	X	X	X	X		
Disinfectants													X								
Driers	X									X											
Fillers	X																				X
Fire extinguishers	X			X			X		X	X										X	X
Fuels																					
Pesticides	X		X			X	X	X	X	X	X	X	X	X							X
Inks	X								X		X			X							X
Lubricants		X		X			X							X							X
Medicine	X		X		X			X	X	X	X	X									X
Oils				X					X		X										
Ointments					X			X	X		X		X								X
Paints	X		X	X			X	X	X	X	X	X	X		X		X		X		X
Photography	X						X			X	X		X			X	X	X			
Pigments	X	X	X	X	X	X	X	X	X	X	X	X	X					X	X		X
Polish	X			X					X								X	X			X
Powders	X									X											X
Preservatives										X	X										X
Suppositories					X																
Water treatment	X								X	X		X									X

other areas of electrical and mechanical engineering. Some 3,400,000 t of aluminum went to these four major areas of usage in the U.S. in 1983.[18] Aluminum foil is widely used for domestic and industrial applications. Indeed, in the U.S. in 1983, approximately 1,800,000 t of packaging material was produced — more than 29% of the total amount of aluminum utilized.[18] Manufacture of domestic appliances, powder, and paste, and metal alloys are other areas of usage.

b. Antimony

The major application of antimony metal used to be in the manufacture of antimonial lead for production of lead-acid type accumulators for motor vehicles.[80] In 1983 this end use accounted for 836 t of antimony in the U.S., compared to 5628, 1318, and 1136 t for manufacture of flame retardants, plastics, and ceramics, respectively.[18] Antimony trioxide is the major compound incorporated into resins as a flame retardant.[80] Minor use of antimony includes the manufacture of ammunition, bearings, pigments, and rubber.

c. Arsenic

It has been estimated that approximately 70% of total arsenic consumption is for the manufacture of agrochemicals such as insecticides and herbicides.[81] Arsenic was at one time incorporated into paint pigments, but this was discontinued when it was found that poisonous arsine and trimethyl arsine gases formed in damp conditions.[82] Wood preservatives, nonferrous alloys, and battery grids contain arsenic,[81] as do fireworks, glass, ceramics, and electronic products such as semiconductors and photoconductors.[79] Heavy industry employs arsenic in the extraction of ores and manufacture of ammonia and sulfuric acid.[78]

d. Beryllium

This metal is used mainly in the manufacture of nuclear reactors, along with other electrical and electronic applications. A total of 181 t was consumed in the U.S. in 1983.[18]

e. Bismuth

Pharmaceuticals manufacture utilized 500.8 t of bismuth in the U.S. in 1983, some 48% of all the metal used. However, this usage and its incorporation into aluminum alloys is on the wane.[83] The metal is also incorporated into low-melting-point and malleable alloys and catalysts.

f. Cadmium

The foremost industrial source of cadmium is the electroplating and coating industry which accounted for approximately 33% of the world's use[84] in 1974 and 1623 t in the U.S. in 1981, or 34% of the total.[18] In pigments for paint is the next most common end use, followed by battery manufacture and in plastics manufacture, especially as a polyvinyl chloride (PVC) stabilizer. Other uses include alloys, solders, fungicides, and chemicals in photography, process engraving, and rubber curing.[79,84]

g. Chromium

This metal is principally used for the formation of alloys that are employed in the transport, building, and machinery-construction industries. Approximately 57% of all chromium in 1983 in the U.S. was used in these industries, a total of some 158,000 t.[18] Nichrome alloys are used in heater elements, high-temperature alloys in industrial equipment, hot part alloys in jet-engine turbines, and "super-stainless" low-iron alloys for hot, corrosive environments.[85] Chromium is also utilized in chemicals manufacture with 33,000 t going to this industry in the U.S. in 1983.[18] It is incorporated into rust inhibitors, inks, pigments, preservatives, and also used in dyeing and tanning.[78,79]

Table 6
LEAD CONSUMPTION BY END USE IN THE U.S. AND U.K. IN 1983[18]

Use	Quantity (10^3 t/year) U.S.	Quantity (10^3 t/year) U.K.
Batteries	537	87
Anti-knock agents	89	54
Pigments and oxides	56	21
Ammunition	40	—
Solder	18	8
Sheet, pipes, shot, etc.	13	88
Alloys	12	20
Cable covering	10	19
Type metal	5	—
Caulking	1	—
Others	300	10
Total	**1081**	**307**

Table 7
MERCURY CONSUMPTION BY END USE IN THE U.S. AND U.K. in 1983[18]

Use	Quantity (t/year) U.S.	Quantity (t/year) U.K.
Electrical	929	74
Chlor-alkali process	278	93
Paints	209	56
Measurement and control	140	37
Dental	55	11
Catalysts	17	—
Laboratories	10	7
Others	58	92
Total	**1696**	**370**

h. Cobalt

Most cobalt is used in the production of alloys — approximately 63% of consumption in the U.S. in 1983.[18] These alloys are for cutting and wear-resisting materials, steels, welding materials, and permanent magnets.[86] Minor uses include chemicals incorporated into paints, pigments, inks, and glass.[79]

i. Copper

The metal itself is widely used in electronics, plating, and brass and other alloys,[78,79] It is also employed in chemicals for fungicides, fertilizers, and rayon manufacture.

j. Lead

The principal use of lead is in the manufacture of batteries (Table 6). Lead is also utilized for anti-knock additions for gasoline, pigments, ammunition, solder, and cable covering.[18] Minor uses include caulking compounds, piping, and type metal. As shown in Table 6, in 1983 approximately 50% of lead in the U.S. and 28% in the U.K. was used in battery manufacture.

k. Manganese

The vast majority of manganese is used in the production of alloys, particularly in combination with aluminum and steel and to a lesser extent copper.[87] Compounds of this metal are used in paint formulations, batteries, glass, oils, disinfectants, fertilizers, and some petrol additives.[79]

l. Mercury

In the U.S. the majority of mercury, some 55% of the total in 1983, is used for the production of electrical equipment, especially lamps (Table 7). The chlor-alkali process for the production of chlorine and caustic soda (sodium hydroxide) also employs large quantities of this metal. Paints, instrumentation for measurement and control, and amalgam in dentistry are also products in which mercury is used. Minor uses of mercury include laboratory work, catalysts, and agrochemicals.

m. Molybdenum

Manufacture of alloys is the largest single use, followed by chemicals and ceramics production.[18,79] Molybdenum is utilized in catalysts, pigments, electrical products (such as contacts and lamp filaments), fertilizers, and lubricants.

n. Nickel

The principal use of nickel is in the production of alloys, accounting for 79% of consumption in the U.S. in 1983 (almost 92,000 t).[18] Electroplating is the second largest use, accounting for approximately 18% of U.S. consumption in 1983. Nickel is also utilized in pigments, catalysts, electrical contacts, and batteries.[79]

o. Selenium

This metal is in widespread use in the manufacture of semiconductors, rectifiers, and photoreceptors for the electrical and electronics industries. Approximately 193 t, or 35% of the total, was consumed for this purpose in the U.S. in 1983.[18] Other major uses of selenium include ceramics, glass, and pigments. The metal and its compounds are also incorporated into alloys, catalysts, deodorants, lubricants, pesticides, and pharmaceuticals. The rubber industry employs selenium for acceleration and vulcanization.

p. Silver

Manufacture of chemicals for photographic products consumes more silver than any other industry. In the U.S. in 1983 almost 44% or 1612 t of all silver was used in this way.[18] A large amount of silver is also used for domestic items such as solid sterling silverware, plated silverware, and jewelery. Other industrial uses of the metals include the production of electrical contacts and switching gear, batteries, catalysts, and mirrors. Some silver is still used in coinage in several countries and also has dental and medical applications.

q. Tellurium

Approximately 75% of world consumption in the mid 1970s was in metallurgical applications, both ferrous and nonferrous.[88] Tellurium is also used in catalysts, photoreceptors, and the rubber industry.

r. Thallium

Industrial uses of thallium are limited but do include some alloys, electronic devices, and special glasses.[89] Compounds of thallium have also been used in synthetic organic chemistry as catalysts.

s. Tin

The metal is principally used in making alloys and plating. Some tin is used for solder. Trialkyl and triaryl tin compounds are used as fungicides and pesticides and also as antimicrobial agents in antifouling paints.[82]

t. Vanadium

Vanadium is primarily used in the manufacture of iron and steel alloys, accounting for approximately 83% of the total consumption of this metal of 2972 t in the U.S. in 1983.[18] Other alloys and superalloys accounted for a further 16% of vanadium consumption. The metal is also used for coloring ceramics and in making oxidation catalysts.[82]

u. Zinc

A great deal of zinc is used in brass and bronze alloys and for galvanization. Compounds of zinc are used in paints, plastics, and rubber, as well as some in cosmetics and pharmaceuticals.[18,79] Industrial applications include batteries, fertilizers, lights, televisions, and tires.

2. Selected Industries

Heavy metals appear in the wastewaters of many industrial plants as a contaminant, as well as from processes in which the metal is used. A good example of this is the power-generation industry which uses the fossil fuels coal and oil and produces wastes which can contain chromium, vanadium, and zinc. Table 8 summarizes the heavy metals that may be present in wastewaters from a variety of industries. Some of these industries are considered in further detail below.

a. Inorganic Chemicals Manufacture

The United States Environmental Protection Agency (EPA) has issued regulations for discharges from inorganic chemicals manufacturing processes.[91] The industries included in this category were the chlor-alkali process and the manufacture of hydrogen fluoride, titanium, dioxide, aluminium fluoride, chrome pigments, sodium dichromate, copper and nickel sulfates, and sodium bisulfite. In each case, at least one of the controlled pollutants (i.e., that present at the highest concentration) was not a primary constituent of any of the raw materials used. During the production of copper sulfate, the raw materials are copper, sulfuric acid, water, and air, yet nickel is regarded as a significant discharge along with copper. Manufacture of aluminum fluoride uses alumina hydrate and gaseous hydrogen fluoride, yet the main pollutants discharged are chromium and nickel. Many other metals are also present in these effluents at low concentrations.

Investigations of marine environments in the vicinity of chlor-alkali plant discharges have indicated severe mercury pollution of sediments and fauna.[92,93] This is unsurprising if effluents from such plants are discharged in an uncontrolled manner, as the wastewaters from electrolysis units can contain up to 35 mg/ℓ mercury.[94] At the same plant in Egypt, the mercury concentration in the final wastewater was approximately 1.1 mg/ℓ, equivalent to an annual mercury emission level of 10,200 kg/year, almost five times the generally accepted level for chlor-alkali process.[95] Chromium, copper, lead, manganese, nickel, tin, and zinc were considered to be present at normal concentrations.[94] Nickel concentrations as high as 890 mg/ℓ have been detected in wastewaters from one nickel sulfate manufacturing plant.[96]

It should be noted that the majority of inorganic chemicals manufacturing plants listed by the EPA discharge directly to water courses, and not to sewers.[91] From this inventory, it is likely that, of this type of manufacturing plant, only nickel sulfate and chrome pigment production could be a major problem where metal inputs to municipal wastewater treatment plants in the U.S. are concerned.

b. Electroplating

The very nature of the electroplating industry means that its wastewaters can be highly contaminated with heavy metals. Chrome plating is a relatively common metal-finishing technique and for this reason effluents from such plants are often investigated.[97-99] A study of a small chrome-plating plant in the U.S. indicated that the nickel/chrome rinse wastewater had concentrations of chromium and nickel as high as 1.74 mg/ℓ and 4.10 mg/ℓ, respectively.[99] Rinse waters from the copper-plating line at the same plant also contained high chromium levels of 0.84 mg/ℓ and a copper concentration of 0.44 mg/ℓ. It has been estimated that there are approximately 11,000 to 14,000 companies in the U.S. with electroplating facilities, many of which are small electronics-related plants.[100] Plants involved in semiconductor manufacture produce effluents high in copper and nickel, arising from rise waters.[101] Manufacture of chips and printed circuit boards also result in discharges with a high copper content, in addition to lead and other heavy metals (Table 9). The EPA has published discharge limits for electroplating facilities which include limitations on metal concentrations.[102]

c. Metal Finishing

The discarded mill solution from gold milling operations is known as the "barren bleed"

Table 8
HEAVY METALS IN WASTEWATERS FROM DIFFERENT CATEGORIES OF INDUSTRIES

	Al	Sb	As	Bi	Cd	Cr	Co	Cu	Fe	Pb	Mn	Hg	Mo	Ni	Se	Ag	Te	Tl	Sn
Metal industries																			
Power plants (steam generation)						X													
Foundries — ferrous		X	X		X	X	X	X	X	X	X		X	X	X		X		X
Foundries — Nonferrous	X	X	X	X	X	X		X		X	X				X	X			
Plating	X				X	X		X						X	X	X		X	
Chemical industries																			
Cement and glass	X		X			X									X			X	
Organics and petrochemicals	X		X		X	X			X	X		X							X
Inorganic chemicals	X		X		X	X			X	X		X							X
Fertilizers	X		X		X	X		X	X	X				X					
Oil refining	X		X		X	X		X	X	X				X					
Others																			
Paper						X		X		X		X		X					
Leather						X													
Textiles		X				X													
Electronics		X						X				X			X	X			X

Adapted from Dean, J. G., Bosqui, F. L., and Lanouette, K. H., *Environ. Sci. Technol.*, 6, 518, 1972.

Table 9
CONCENTRATIONS OF HEAVY METALS IN WASTEWATERS FROM SMALL ELECTROPLATING PLANTS[100]

Plant	Flow (m³/day)	Metal concentration (mg/ℓ)				
		Cu	Fe	Pb	Ni	Sn
Chips	2.3—9.1	60	—	1.2	1.9	<0.43
Printed circuits	0.5—13.6	—	40	0.1	—	1.0
Printed circuits	230	85	—	0.8	—	0.3

and contains high concentrations of iron cyanide — the iron content varying from 25 to 30 mg/ℓ.[103] However, concentrations of copper and zinc in the untreated effluent are much higher, at between 180 and 210 mg/ℓ copper and between 60 and 68 mg/ℓ zinc. Upon treatment by ion-exchange techniques, the effluent can be made suitable for discharge to sewer main since it contains only <0.5 mg/ℓ copper and 0.2 mg/ℓ zinc.[103]

Carter et al.[104] studied a lagoon treatment system that received wastewaters from a piston-ring manufacturing plant in the U.S. The influent to the lagoon, i.e., raw wastewater from the plant, in 1980 contained 0.04 to 0.83 mg/ℓ lead and 0.04 to 0.74 mg/ℓ zinc. Such high concentrations of lead and zinc would have exceeded the permitted levels for discharge to sewer 80 and 20% of the time, respectively. Treatment in the lagoon reduced the concentration of pollutants to within permitted levels. However, lead concentrations discharged to the municipal sewer in the first three quarters of 1980 exceeded the permitted 0.05 mg/ℓ; during the first quarter it was as high as 0.18 mg/ℓ, three times the limit. Hexavalent chromium concentrations also exceeded limitations ranging up to 0.77 mg/ℓ. Therefore the loading from such an industrial facility to a sewerage system can be considerable even with some kind of pretreatment.

In the U.S. the aluminum-forming industry includes 279 plants with rolling, drawing, extruding, and forging processes along with support operations.[105] The wastewaters from the industry can be classified into four main groups: oil-in-water emulsions, free oils, process-cooling wastewaters, and cleaning wastewaters, with the latter two contributing most of the heavy metals. Indeed, very high concentrations of some metals were found in cooling wastewater; e.g., 46 mg/ℓ chromium, 17 mg/ℓ lead, and 5.2 mg/ℓ zinc.[105] Summarizing the wastewater contents from aluminum-forming plants, Carlton et al.[105] found median concentrations of approximately 20, 0.14, 0.16, 0.18, and 0.02 mg/ℓ of aluminum, chromium, copper, lead, nickel, and zinc, respectively. Peak concentrations detected were >100 mg/ℓ for each metal. Aluminum-finishing wastewater can also contain high concentrations of nickel, up to 2.2 mg/ℓ, in addition to high levels of aluminum itself.[106]

d. Photoprocessing

Lytle[107] studied the fate and speciation of silver in three categories of publicly owned wastewater-treatment plants. Silver concentrations in the influent to plants receiving essentially domestic sewage were from 5 to 10 μg/ℓ, and in those receiving discharges from industrial silver users, from 8 to 30 μg/ℓ. Wastewaters with a substantial contribution from photoprocessing discharges had from 460 to 2200 μg/ℓ silver. In the U.S. in 1982, approximately 43% of all silver was used in the photographic industry,[108] so this is the most important silver source in wastewaters.

e. Paints, Inks, and Pigments

A comprehensive sampling program of untreated wastewater from the paint industry in the U.S. was undertaken in 1977/1978.[109] Of the 20 heavy metals analyzed, median concentrations detected were high for aluminum at 100 mg/ℓ, lead at 0.8 mg/ℓ, and zinc at 10 mg/ℓ. However, maximum concentrations observed exceeded the median levels by several times; e.g., cobalt at 40 mg/ℓ, lead at 62 mg/ℓ, and zinc at 900 mg/ℓ. The high metal concentrations reflect their usage in pigments for paints. Berlow et al.[109] also surveyed wastewaters from ink-manufacturing plants in the U.S. and found that significant median concentrations of aluminum at 20 mg/ℓ, molybdenum at 1.0 mg/ℓ, chromium at 20 mg/ℓ, and lead at 50 mg/ℓ were present. Maximum detected lead and molybdenum concentrations were as high as 900 and 600 mg/ℓ, respectively. In the U.S. the paint industry is comprised of approximately 1500 plants, and the ink industry, about 500 sites which discharge approximately 3000 m^3/day of wastewater.[109]

Shaul et al.[110] studied the effectiveness of a powdered activated carbon addition to activated sludge in the treatment of wastewaters from a dye and pigment processing factory. Concentrations of metals during two separate periods of sampling ranged from 0.6 to 13.9 mg/ℓ chromium, 0.2 to 2.4 mg/ℓ copper, 1.5 to 13.6 mg/ℓ lead, and 0.3 to 12.2 mg/ℓ zinc. The wastewaters from the plant were discharged directly to the sewerage system.

III. HEAVY METALS IN WASTEWATER

A. Source Contributions

Several studies have investigated the contributions of the three major sources — runoff, domestic, and industrial — to the overall metal loadings at wastewater-treatment plants. An investigation in New York City indicated that, in general, industrial sources accounted for more than half the chromium and nickel received at the treatment works (Table 10).[111] Of the 65% contribution from industrial nickel sources, 62% derived from electroplating operations alone. Almost half the cadmium and zinc were from residential areas as well as approximately two thirds of copper, which included 20% accounted for in the water supply. Runoff contributed only about 10% of all metals with the exception of zinc which was greater probably due to its deposition on roads. Another study in New York by Davis and Jacknow[112] suggested that the contribution of residential areas was lower for cadmium, chromium, copper, and zinc (Table 10). Only domestic nickel sources at 34% were estimated to be greater than the figure of 25% quoted by Klein et al.[111] Davis and Jacknow[112] also found a high nickel contribution from electroplating — some 53% of the total. These authors also investigated metal loadings in the cities of Muncie, Ind. and Pittsburgh, Penn. Industrial sites in Muncie were easily the major source of metals with >90% for chromium and lead (Table 10). Therefore metal loadings from residential areas were low, ranging from >3% for chromium to a maximum of one third for copper. Residential loadings in Pittsburgh were considered reasonable, with the exception of a high value for copper (Table 10). The authors also calculated the residential loading factors for each city and metal, which ranged from 2.7 g/day/1000 capita for cadmium in Muncie to 95 g/day/1000 capita for zinc in Muncie and New York, and demonstrated that, in general, Muncie and Pittsburgh were similar, with higher values for New York, probably due to some light industrial and commercial discharge being included in the Bowery Bay residential area sampling program. During a survey of the sewerage system of Kokomo, Ind., Yost and Wukasch[113] found that a large proportion of most metals received at the municipal treatment works could be accounted for by discharges from only 12 point sources (Table 10). However, cadmium was a notable exception and estimated loadings from residential areas were so low that the total received at the works could not be accounted for. Koch et al.[114] estimated the percentage of industrial and domestic heavy metals contributed from industrial sources for two districts in Camden County, N.J. and found that, in one district, only cadmium was largely of industrial origin, whereas all seven metals analyzed were principally from industrial sites in the second district (Table

Table 10
PERCENTAGE CONTRIBUTIONS OF DIFFERENT SOURCES OF HEAVY METALS IN THE INFLUENT TO WASTEWATER-TREATMENT WORKS

Location	Source	Metals (%)						Ref.
		Cd	Cr	Cu	Pb	Ni	Zn	
New York City	Electroplating	33	43	12	—	62	13	111
	Other industries	6	9	7	—	3	7	
	Runoff	12	9	14	—	10	31	
	Residential[a]	49	28	67	—	25	49	
New York City	Electroplating	19	14	15	—	53	7	112
	Industries, runoff	43	58	46	—	12	76	
	Residential	38	27	38	—	34	16	
Pittsburgh, Penna.	Residential	63	23	96	63	19	32	112
Muncie, Ind.,	Industrial	—	98	67	91	89	85	112
	Residential	—	3	36	10	13	17	
Kokomo, Ind.	Point source[b]	29	86	98	115	92	204	113
	Residential[b]	<1	<1	5	17	6	2	
Camden County, N.J.								114
District I	Industrial[c]	79	23	10	19	29	12	
District II	Industrial[c]	30	77	28	29	62	47	
Columbus, Ohio								115
Jackson Pike	Water supply	0	<1	10	<1	1	27	
	Residual	6	1	11	6	4	8	
	Runoff[d]	0	5	3	36	1	3	
	Industrial	37	14	14	12	17	35	
Southerly	Water supply	0	<1	18	<1	3	45	115
	Residential	9	1	21	11	12	13	
	Runoff[d]	0	4	3	33	2	2	
	Industrial	118	40	60	41	57	37	
Maple Lodge, U.K.	Domestic	29	15	57	69	29	58	116

[a] Residential includes 20% Cu and 7% Zn from water supply.
[b] Lack of "balances" due to nonsimultaneous sampling.
[c] Percentage of industrial and domestic total.
[d] Includes infiltration and inflow.

10). Stonebrook et al.[115] observed that much of the influent cadmium, chromium, copper, lead, and nickel at the Jackson Pike wastewater-treatment works, Columbus, Ohio, was unaccounted for in the estimated collection system metals loading, whereas influent metals at the Southerly plant could be accounted for (Table 10). Again industrial sources contributed the major portions of most metals, although runoff was of significance for lead.

Several studies have demonstrated that significant heavy-metal discharges can be associated with residential areas. Callahan et al.[117] observed that copper, chromium, lead, manganese, nickel, and zinc loadings from old residential areas in a city in the U.S. were in general greater than those from new residential areas, especially in the case of zinc. In a survey of heavy metals in sewage sludge in West Germany, high zinc concentrations were attributed to the use of galvanized pipes in domestic plumbing.[118] During another survey of municipal sludges in the U.S., a very high copper content was observed for a small community with almost entirely domestic discharges.[119] It was noted that the copper concentration was so high that it exceeded values recommended for disposal of sludge to land.[120] Brindley et al.[116] estimated the percentage domestic contribution to the heavy-metal loading at the Maple Lodge Sewage Treatment Works (Thames Water Authority, U.K.) and found that more than half the copper, lead, and zinc was attributed to this source (Table 10). The authors also suggested that this was due to the use of these metals in domestic plumbing.

Brindley et al.[116] also observed the influent concentrations of metals between 1974 and 1979 and noted that changes in sewer-discharge consent standards reduced the loadings of cadmium, chromium, lead, and nickel but had little effect on copper and zinc, presumably as they were of domestic origin. The sewer-discharge consents were cadmium, 2 mg/ℓ; copper, 5 mg/ℓ; chromium, 5 mg/ℓ; lead, 5 mg/ℓ; mercury, 1 mg/ℓ; nickel, 5 mg/ℓ; silver, 2 mg/ℓ; and zinc, 10 mg/ℓ. The maximum permissible flow of industrial wastewaters discharged to sewers was 43,780 m^3/day, of which 14.6% was from 22 photoprocessing laboratories, 6.6% from 33 electroplating plants, and 4.5% from 17 printing facilities. In a different region of the Thames Water Authority area, the Metropolitan Public Health Division, sewer-discharge or trade-effluent consents were cadmium 1, mg/ℓ; chromium, 15 mg/ℓ; copper, 5 mg/ℓ; lead, 10 mg/ℓ; nickel, 5 mg/ℓ, and zinc, 5 mg/ℓ.[121] The total concentrations of several metals received at the Mogden sewage-treatment works fell between 1973 and 1979/1980 due, according to the authors, to the reduction and enforcement of the industrial-discharge consents.[121] Cadmium loadings decreased by 54% from 11.6 to 5.2 kg/day, copper by 30% from 75 to 107 kg/day, nickel by only 12% from 78 to 69 kg/day, and zinc by 45% from 297 to 163 kg/day. These figures demonstrate that the nickel loading was either mainly from domestic sources, which would seem to be unlikely, or that the change in consents had little effect on those industries discharging the metal. Dakers and King[121] also estimated that the industrial-discharge contributions to cadmium loadings at Beckton, Crossness, and Riverside sewage-treatment works were 61, 70, and 77%, respectively.

In the Severn-Trent Water Authority area in the U.K., there are 6532 industrial sites discharging directly to sewers with a further 1020 effluenting to water courses.[122] There were a further 6500 point source discharges of process waters to the sewerage system. However, Harkness[122] noted that only one or two treatment works received >30% industrial flow and the majority were <20%. A survey of the influent metals concentrations for 239 wastewater-treatment plants in the U.S. divided the results into two categories — those with industrial discharges contributing >4% of the overall influent flow and those with <4% derived from industrial sources.[122,123] Aluminum, arsenic, cadmium, chromium, copper, lead, manganese, mercury, nickel, and zinc were all present at higher concentrations in the raw sewages with >4% industrial flow, and chromium, copper, lead, manganese, and zinc were an order of magnitude greater than the <4% industrial-derived influent.

B. Case Studies
1. Camden County, N.J.

Koch et al.[114] undertook a study of the sources of heavy metals to wastewater-treatment plants in Camden County, N.J. in order to assess the impact of pretreatment on the heavy-metal concentration in sludge for disposal. The Delaware Basin area of Camden County was served by 40 existing wastewater-treatment works. It was planned to replace these works with two regional facilities and therefore the study of metal sources was undertaken in two districts designated I and II.

As industrial discharges were reckoned to contribute about 20% of the total wastewater flow in the districts as a whole, two methods were combined to estimate the heavy-metal load from industrial activities. Approximately 1300 industrial concerns were identified by their U.S. Department of Commerce Standard Industrial Classification Number and were sent questionnaires requesting information on location, products, water use, effluents discharged, and other facts. Fifty percent of questionnaires were returned covering the industrial groups automotive, personal servies, service stations, electrical, fabricated and primary metal products, rubbers and plastics, petroleum refining, chemicals, printing, and paper and food processing. This information was combined with the results of a sampling survey of 106 industries and 40 publicly owned treatment works to construct a mass balance for heavy

Table 11
PERCENTAGE OF HEAVY METALS AND INDUSTRIAL FLOWS CONTRIBUTED BY STANDARD INDUSTRIAL CLASSIFICATION (SIC) GROUPS

SIC group	Metal (%)							
	Cd	Cr	Cu	Pb	Hg	Ni	Zn	Flow
Personal services	59	2	13	15	32	5	12	1
Durable goods	<1	9	<1	7	2	22	2	<1
Electrical/electronics	<1	3	17	3	3	<1	2	6
Fabricated metals	25	4	13	12	—	27	5	<1
Primary metals	<1	39	1	<1	9	1	2	4
Chemicals	<1	10	19	10	4	7	19	16
Paper	<1	3	—	9	9	4	5	14
Apparel	<1	—	<1	<1	—	—	27	<1
Food	3	14	11	25	25	12	17	51

From Koch, C. M., Stroka, J. G., Perna, R. K., and Foerster, R. E., *J. Water Pollut. Control Fed.*, 54, 339, 1982. With permission.

metals in Districts I and II. The metals analyzed were cadmium, chromium, copper, lead, mercury, nickel, and zinc. The mass balance assumed that metals were from three main sources: industrial, domestic, and storm water. A limited sampling study demonstrated that storm runoff contributed little to the heavy-metals load of the Delaware Basin, in contrast to other studies of storm water contributions to the metal loading of combined sewerage systems.[47] Domestic wastewater heavy-metal concentrations were assessed by sampling raw sewage at six treatment plants that received almost exclusively domestic flows. A typical domestic metal concentration for use in the mass balance was derived from this data and from values reported in the literature.

Greater than 70% of the metals appearing in the influent at wastewater-treatment plants could be accounted for by the domestic and industrial contributions with the rest unspecified. The proportion of industry-derived metals to the combined industrial and domestic load was also estimated for both districts, as shown in Table 10. This table indicates that in District I, all metals, with the exception of cadmium, were largely domestic in origin. In District II, a higher proportion of the metals, again with cadmium the exception, were derived from industrial sources. Koch et al.[114] also estimated the proportions of industrial flows and heavy metals by types of industries, as shown in Table 11. These data demonstrate that the most important source for any particular metal can vary considerably. Most cadmium and mercury were derived from personal services; food industries contributed a quarter of the total industrially derived lead; copper came mainly from electronics, zinc from clothing manufacture, and chromium and nickel from metals industries. More than half of the industrial flow in the Delaware Basin was derived from food-related industries and therefore contributed a considerable proportion of all metals to the total from industrial sources, despite the fact that the actual metal content of food-industry wastes was quite low. The authors concluded that pretreatment of industrial discharges in District I would reduce the loading of cadmium by 70%. Nickel would be the next most affected metal, but the reduction would be only 10%. In District II, zinc loading would be reduced most if industrial pretreatment was introduced, but only by 27%. The 70% reduction in cadmium from industrial sources would, however, only result in a 10% increase in the quantities of sludge that would be permitted for disposal onto land, so the limiting metal would then be copper.

This study illustrates four major points concerning the sources of heavy metals entering a wastewater-treatment works. In the first instance, the results demonstrate that the contribution of heavy metals from runoff due to storms can be negligible. It should be noted, however, that sampling of storm water runoff in sewers during the study was limited and that this conclusion should be viewed as tentative only. The second fact illustrated is that the contribution of industrial discharges to the overall heavy-metal load is highly dependent on the locality served, as shown by the differences between Districts I and II. Perhaps the most striking conclusion to be drawn from the paper by Koch et al.[114] is that the major contribution of heavy metals does not necessarily originate from those industries discharging wastes with high concentrations of these heavy metals, but from industrial effluents of a high volumetric flow but relatively low concentrations of metals. In the present study, food industries contributed 51% of the volumetric flow of industrial discharges and therefore a large proportion of the industrially derived metals loading. The fourth fact that this study demonstrates is that a large reduction in the contribution of heavy metals from point sources does not necessarily translate into an equivalent increase in the amount of sludge that can be legally applied to land.

2. Kokomo, Ind.

The sources and flow through the sewerage network of the heavy metals cadmium, chromium, copper, lead, nickel, and zinc in the city of Kokomo, Ind. were studied in detail by Yost and Wukasch.[113] Kokomo has a population of 42,000 with a wide variety of industries and well-defined residential areas served by a combined sanitary and storm sewer system. The trunk sewers serving the publicly owned treatment works, in addition to twelve major point sources, were sampled in order to determine the concentrations of heavy metals. The point sources were sampled at 2-hr intervals for 24 hr over a consecutive 3-day period, as were the trunkline sewers, over a 3-month period in 1979. At the same time, flows of the industrial discharges and in-the-trunkline sewers were determined.

Point Sources 1, 2, and 3 were all situated on the New Pete's Run trunk sewer. The point source designated as 1 was a manufacturer of transmission and die-cast materials with effluent pretreatment consisting of batch settlement only, with the supernatant being discharged to the sewerage system. Zinc was the most significant metal in this effluent. Point Source 2 discharges originated from a factory involved in the plating of circuit boards; therefore copper was the major metal constituent of the effluent. At the site of Point Source 3, radio and semiconductor manufacture was undertaken, and the discharge contained large quantities of cadmium, contributing 0.32 kg/day or >50% of the measured cadmium point source loading to the Kokomo works. Point Source 4 derived from a vehicle frame component factory and the effluent received some batch treatment, as did the first three sources, prior to discharge to the sewer. The electroplating plant at Point Source 5 produced aqueous waste contributing 0.26, 37.2, and 2.2 kg/day of cadmium, chromium, and nickel which represented 41, 84, and 31%, respectively, of the point-source derived metals entering the sewerage system. There was no pretreatment at this site, which accounts for the high metal discharges. Site 6 produced effluent from a plant manufacturing galvanized metals — thus accounting for the 85% contribution to the industrial loading of zinc. This plant was the only point source to practice continuous pretreatment. Point Sources 7 and 8 produced galvanized fencing and aluminium products, respectively. Alloys containing cobalt, iron, and nickel were formed and forged at the site of Point Source 9. Incomplete pretreatment of a nickel-based alloy waste probably accounted for the high nickel discharge of 3.8 kg/day. The alloy fabrication plant and laundry at Sources 10 and 11, respectively, contributed negligible amounts of metals to the total industrial loading. A significant quantity of copper, at 5.2 kg/day (almost 50% of the total), was discharged by the printing plant sited at Point Source 12. In addition, high amounts of chromium were detected, about 2.2 kg/day, due to the use of copper/chromium plate rolls in the process.

Concentrations of metals from the three trunkline sewers which did not have any point

Table 12
SUMMARY OF METAL INPUTS TO THE KOKOMO TREATMENT PLANT

Metal	Residential (kg/day)	Point source (kg/day)	Total residential and point source (kg/day)	Received at wastewater plant (kg/day)
Cd	0.01	0.63	0.64	4.74
Cr	0.07	44.4	44.5	114.0
Cu	0.59	10.8	11.4	24.3
Ni	0.43	6.9	7.3	16.7
Pb	0.10	0.68	0.78	1.3
Zn	2.56	277.6	280.2	300.0

From Yost, K. J. and Wukasch, R. F., *Environ. Monit. Assess.*, 3, 61, 1983. With permission.

sources discharging to them were determined in order to estimate the residential loading to the wastewater-treatment works. As the sampling scheme was not completely comprehensive, a "random superposition" method was used to determine whether trunkline metal flows could be accounted for by the point source discharges. Analysis of metals in New Pete's Run suggested the presence of a further unidentified point source discharging cadmium, chromium, and nickel to the sewer. Only the quantities of cadmium detected in the Washington feeder were not attributable to the effluents from Point Sources 4, 5, and 7. Zinc and chromium loadings in the Pete's Run sewer were much higher than those discharged at Point Sources 9 and 10. All other heavy metals were accounted for by these two point sources. All of the South Northside Interceptor flows were attributable to wastewaters from Point Sources 8 and 12, with the notable exception of zinc. However, discharges of zinc from the factory at Point Source 8 were found to be highly variable, and a batch flow from this site during the trunkline sewer sampling program may account for this observation.

Comparison of the quantities of metals derived from the point source and residential sources with the detected metal concentrations at the treatment works influent demonstrated, in the opinion of the authors, that all metals, except for cadmium, could be accounted for, as shown in Table 12. Less than 15% of the influent cadmium could be attributed to the amounts detected in the sewerage system. For all other metals, approximately 40% or greater of the loading received at the treatment plant could be accounted for by the total point and residential source flows. It is also evident that the overwhelming majority of the heavy-metal loadings came from 12 point sources investigated. This information was used to reduce the industrially derived metal sources to ensure that Kokomo sludge complied with land-application regulations.

REFERENCES

1. **Chalmers, R. A.**, Editor's footnote; in **Florence, T. M.**, The speciation of trace elements in waters, *Talanta*, 29, 345, 1982.
2. **Parker, S. P., Ed.**, *Dictionary of Scientific and Technical Terms*, 3rd ed., McGraw-Hill, New York, 1984.
3. **Waldron, H. A. and Stöfen, D.**, *Sub-Clinical Lead Poisoning*, Academic Press, London, 1974, chap. 1.
4. Organization for Economic Co-operation and Development, Mercury and the Environment, O.E.C.D., Paris, 1974.
5. **Imboden, D. M. and Stumm, W.**, Influence of man on the geochemical circulation of the atmosphere, *Chimia*, 27, 155, 1973.
6. **Nriagu, J. O.**, Global inventory of natural and anthropogenic emissions of trace metals to the atmosphere, *Nature (London)*, 279, 409, 1979.
7. **Stumm, W. and Bilinski, H.**, Trace metals in natural waters: difficulties of interpretation arising from our ignorance on their speciation, in *Proc. Int. Conf. of Advances in Water Pollution Research*, Jenkins, S. H., Ed., Pergamon Press, Oxford, 1973, 39.
8. **Habib, S. and Minski, M. J.**, Incidence and variability of some elements in the non tidal region of the River Thames, and River Kennet, U.K., *Sci. Total Environ.*, 22, 253, 1982.
9. **Demayo, A., Taylor, M. C., and Taylor, K. W.**, Effects of copper on humans, laboratory and farm animals, terrestrial plants and aquatic life, *Crit. Rev. Environ. Control*, 12, 183, 1982.
10. **Taylor, M. C., Demayo, A., and Taylor, K. W.**, Effects of zinc on humans, laboratory and farm animals, terrestrial plants, and freshwater aquatic life, *Crit. Rev. Environ. Control*, 12, 113, 1982.
11. **Demayo, A., Taylor, M. C., Taylor, K. W., and Hodson, P. V.**, Toxic effects of lead and lead compounds on human health, aquatic life, wildlife, plants, and livestock, *Crit. Rev. Environ. Control*, 12, 257, 1982.
12. **Hutton, M.**, Sources of cadmium in the environment, *Ecotoxicol. Environ. Saf.*, 7, 9, 1983.
13. **Rentsch, J.**, Cadmium in the environment, *Proc. Symp. Cadmium Research in Switzerland*, University of Fribourg, March 19, 1984.
14. **Rösler, H. J. and Lange, H.**, *Geochemical Tables*, Elsevier, Amsterdam, 1972, 231.
15. **Taylor, S. R.**, Trace element abundances and the chondritic earth model, *Geochim. Cosmochim. Acta*, 28, 1989, 1964.
16. **Settle, D. M. and Patterson, C. C.**, Lead in albacore guide to land pollution in Americans, *Science*, 207, 1167, 1980.
17. **Anon.**, *Metal Bulletin Handbook, Volume 2, Statistics and Memoranda*, Metal Bulletin Books, London, 1984.
18. **Anon.**, *Roskill's Metals Databook*, 6th ed., Roskill Information Services, London, 1985.
19. **Stephenson, T. and Lester, J. N.**, Heavy metal removal during the activated sludge process I exent of soluble and insoluble metal removal, *Sci. Total Environ.*, in press.
20. **Ericksson, E.**, The yearly circulation of chloride and sulfur in nature; meterological, geochemical and pedological implications. II, *Tellus*, 12, 63, 1960.
21. **Turekian, K. K. and Wedepohl, K. H.**, Distribution of the elements in some major units of the earth's crust, *Geol. Soc. Am. Bull.*, 72, 175, 1961.
22. **Underwood, E. J.**, Environmental sources of heavy metals and their toxicity to man and animals, *Prog. Water Technol.*, 11, 33, 1979.
23. **Metcalf & Eddy Inc.**, *Wastewater Engineering: Collection, Treatment, Disposal, and Reuse*, 2nd ed., Tchobanoglous, G., Ed., McGraw-Hill, New York, 1978.
24. **Galloway, J. N. and Likens, G. E.**, Calibration of collection procedures for the determination of precipitation chemistry, *Water Air Soil Pollut.*, 6, 241, 1976.
25. **Galloway, J. N., Thornton, J. D., Norton, S. A., Volchok, H. L., and McLean, R. A. N.**, Trace metals in atmospheric deposition: a review and assessment, *Atmos. Environ.*, 16, 177, 1982.
26. **Weiss, H., Bertine, K., Koide, M., and Goldberg, E. D.**, The chemical composition of a Greenland glacier, *Geochim. Cosmochim. Acta*, 39, 1, 1975.
27. **Herron, M. M., Langway, C. C., Weiss, H. V., and Cragin, J.**, Atmospheric trace metals and sulfate in the Greenland ice sheet, *Geochim. Cosmochim. Acta*, 41, 915, 1977.
28. **Boutron, C. and Lorius, C.**, Trace metals in Antarctic snows since 1914, *Nature (London)*, 277, 551, 1979.
29. **Cawse, P. A.**, A Survey of Atmospheric Trace Elements in the U.K.: Results for 1979, United Kingdom Atomic Energy Authority, AERE R9886, Harwell, U.K., 1981.
30. **Lindberg, S. E. and Turner, R. R.**, Trace metals in rain at forested sites in the eastern United States, *Proc. 4th Int. Conf. Heavy Metals in the Environment*, CEP, Edinburgh, 1983, 107.
31. **Navarre, J. L., Priest, P., and Ronneau, C.**, Mid distance transfer of heavy metals from industrial sources to a rural environment, *Proc. 2nd Int. Conf. Heavy Metals in the Environment*, CEP, Edinburgh, 1979, 279.

32. **Vouk, V. B. and Piver, W. T.**, Metallic elements in fossil fuel combustion products: amounts and forms of emissions and evaluation of carcinogenicity and mutagenicity, *Environ. Health Perspec.*, 47, 201, 1983.
33. **Müller, J.**, Beryllium, cobalt, chromium and nickel in particulate matter of ambient air, *Proc. 2nd Int. Conf. Heavy Metals in the Environment*, CEP, Edinburgh, 1979, 300.
34. **Harrison, R. M. and Williams, C. R.**, Atmospheric cadmium pollution at rural and urban sites in north-west England, *Proc. 2nd Int. Conf. Heavy Metals in the Environment*, CEP, Edinburgh, 1979, 262.
35. **Breder, R., Flucht, R., Ferrara, R., Barghigiani, C., and Seritti, A.**, Mercury levels in the air of a Mediterranean area, *Proc. 4th Int. Conf. Heavy Metals in the Environment*, CEP, Edinburgh, 1983, 151.
36. **Nürnberg, H. W., Valenta, P., and Nguyen, V. D.**, The wet deposition of heavy metals from the atmosphere in the Federal Republic of Germany, *Proc. 4th Int. Conf. Heavy Metals in the Environment*, CEP, Edinburgh, 1983, 115.
37. **Kwapulinski, J. and Sarosiek, J.**, Migration of beryllium in an industrial region, *Proc. Int. Conf. Environmental Contamination*, CEP, Edinburgh, 1984, 532.
38. **Bengtsson, S. and Tyler, G.**, Vanadium in the Environment, Monitoring Assessment Research Centre Report No. 2, Monitoring Assessment Research Center, London, 1976.
39. **Dewling, R. T., Trichon, M., Chassanoff, E., Rothstein, R., and O'Sullivan, W.**, Impact of heavy metal emissions in the New York - New Jersey metro area, *Proc. 3rd Int. Conf. Heavy Metals in the Environment*, CEP, Edinburgh, 1981, 52.
40. **Harris, M. R.**, Concentration of lead in the atmosphere and soil measured in the vicinity of a secondary lead smelter, *Environ. Technol. Lett.*, 2, 233, 1981.
41. **Thompson, L. H. and Eves, E. G.**, Hogsmill Valley Sewage Treatment Works, Thames Water Authority, London, 1978.
42. **Page, A. L. and Ganje, T. O.**, Accumulations of lead in soils for regions of high and low motor vehicle traffic density, *Environ. Sci. Technol.*, 4, 140, 1970.
43. **Peyton, T., McIntosh, A., Anderson, V., and Yost, K.**, Aerial input of heavy metals into an aquatic ecosystem, *Water Air Soil Pollut.*, 5, 443, 1976.
44. **Lester, J. N.**, Significance and behaviour of heavy metals in waste water treatment processes. I. Sewage treatment and effluent discharge, *Sci. Total Environ.*, 30, 1, 1983.
45. **Newton, C. D., Shephard, W. W., and Coleman, M. S.**, Street runoff as a source of lead pollution, *J. Water Pollut. Control Fed.*, 46, 999, 1974.
46. **Cawse, P. A.**, Trace Elements in Soil, Vegetables and Rainwater in Walsall in 1978, United Kingdom Atomic Energy Authority, AERE G 1343, Didcot, U.K., 1980.
47. **Roberts, P., Hegi, H. R., Webber, A., and Krädhenbähl, H. R.**, Metals in municipal waste water and their elimination in sewage treatment, *Prog. Water Technol.*, 8, 301, 1977.
48. **Sonzogni, W. C., Chesters, G., Coote, D. R., Jeffs, D. N., Konrad, J. C., Ostry, R. C., and Robinson, J. B.**, Pollution from land runoff, *Environ. Sci. Technol.*, 14, 148, 1980.
49. **Pope, W., Graham, N. J. D., Young, R. J., and Perry, R.**, Urban runoff from a road surface - a water quality study, *Prog. Water Technol.*, 10, 533, 1978.
50. **U.K. Ministry of Health**, Final Report of the Departmental Committee on Ethyl Petrol, His Majesty's Stationary Office, London, 1930.
51. **Dunn, J. T. and Blaxham, H. C. L.**, The occurrence of lead, copper, zinc and arsenic compounds in atmospheric dusts and sources of these impurities, *J. Soc. Chem. Ind. (London)*, 52, 189T, 1933.
52. **Thomas, H. F., Elwood, P. C., Welsby, E., and St. Ledger, A. S.**, Relationship of blood lead in women and children to domestic water lead, *Nature (London)*, 282, 712, 1979.
53. **Thomas, H. F., Elwood, P. C., Toothill, C., and Morton, M.**, Blood and water lead in a hard water area, *Lancet*, 1, 1047, 1981.
54. **Pocock, S. J.**, Factors influencing household water lead: a British national survey, *Arch. Environ. Health*, 35, 45, 1980.
55. **Laxen, D. P. H. and Harrison, R. M.**, The highway as a source of water pollution — an appraisal with the heavy metal lead, *Water Res.*, 11, 1, 1977.
56. **Shaheen, D. G.**, Contributions of Urban Roadway Usage to Water Pollution, Report 600/2-75-004, Environmental Protection Agency, Washington, D.C., 1975.
57. **Hedley, G. and Lockley, J. C.**, Quality of water discharged from an urban motorway, *Water Pollut. Control*, 74, 659, 1975.
58. **Crosby, D. G.**, Environmental chemistry: an overview, *Environ. Toxicol. Chem.*, 1, 1, 1982.
59. **Colwill, S. M., Peters, C. J., and Perry, R.**, Water Quality of Motorway Runoff, Supplementary Report 823, U.K. Transport and Road Research Laboratory, Crowthorne, Berks, U.K., 1984.
60. **Colwill, D. M. and Hickman, A. J.**, Measurement of Particulate Lead on the M4 Motorway at Harlington, Third Report, Research Report 972, U.K. Transport and Road Research Laboratory, Crowthorne, Berks, U.K., 1981.

61. **Bevan, M. G., Colwill, D. M., and Hogbin, L. E.**, Measurements of Particulate Lead on the M4 Motorway at Harlington, Research Report 626, U.K. Transport and Road Research Laboratory, Crothorne, Berks, U.K., 1974.
62. **Colwill, D. M., Peters, C. J., and Perry, R.**, Motorway runoff — the Effects of Drainage Systems on Water Quality, Research Report 37, U.K. Transport and Road Research Laboratory, Crowthorne, Berks, U.K., 1985.
63. **Roberts, P. V., Dauber, L., Novak, B., and Zobrist, J.**, Pollutant loadings in urban storm water, *Prog. Water Technol.*, 8, 93, 1977.
64. **Helsel, D. R., Kim, J. I., Grissard, T. J., Randall, C. W., and Hoehn, R. C.**, Land use influences on metals in storm drainage, *J. Water Pollut. Control Fed.*, 51, 709, 1979.
65. **Morrison, G. M. P., Revitt, D. M., Ellis, J. B., Svensson, G., and Balmer, P.**, Variations of dissolved and suspended solid heavy metals through an urban hydrograph, *Environ. Technol. Lett.*, 5, 313, 1984.
66. **Melonen, M. J.**, Collection and analysis of urban runoff data in Finland, *Water Sci. Technol.*, 17, 1175, 1985.
67. **Harrison, R. M. and Wilson, S. J.**, The chemical composition of highway drainage waters. I. Major ions and selected trace metals, *Sci. Total Environ.*, 43, 63, 1985.
68. **Harrison, R. M., Laxen, D. P. H., and Wilson, S. J.**, Chemical associations of lead, cadmium, copper and zinc in street dusts and roadside soils, *Environ. Sci. Technol.*, 15, 1378, 1981.
69. **Harrison, R. M. and Wilson, S. J.**, The chemical composition of highway drainage waters. II. Chemical associations of metals in the suspended sediment, *Sci. Total Environ.*, 43, 79, 1985.
70. **Clark, A. I., McIntyre, A. E., Lester, J. N., and Perry, R.**, Ambient air measurements of aromatic and halogenated hydrocarbons at urban, rural and motorway locations, *Sci. Total Environ.*, 39, 265, 1984.
71. **Revitt, D. M. and Ellis, J. B.**, Rain water leachates of heavy metals in road surface sediment, *Water Res.*, 14, 1403, 1980.
72. **Harrison, R. M. and Wilson, S. J.**, The chemical composition of highway drainage waters. III. Runoff water metal speciation characteristics, *Sci. Total Environ.*, 43, 89, 1985.
73. **Spector, W. S.**, *Handbook of Biological Data*, W.B. Saunders, London, 1956.
74. **Davis, R. D. and Coker, E. G.**, Cadmium in Agriculture with Special Reference to the Application of Sewage Sludge on Land, Tech. Rep. 139, Water Research Centre, Stevenage, U.K., 1980.
75. **Matthews, P. J.**, Sewage sludge utilization — management as part of water quality control, *Environ. Technol. Lett.*, 1, 65, 1980.
76. **Atkins, E. D. and Hawley, J. R.**, Sources of Metals and Metal Levels in Municipal Waste Waters, Research Report 80, Environment Canada, Ottawa, 1978.
77. **Harr, J. R.**, Biological effects of selenium, in *Toxicity of Heavy Metals in the Environment*, Oehme, F. W., Ed., Marcel Dekker, New York, 1978, 396.
78. **Patterson, J. W.**, Industry sources and control of metals, *Proc. 2nd Int. Conf. Heavy Metals in the Environment*, CEP, Edinburgh, 1979, 617.
79. **Matthews, P. J.**, Control of metal application rates from sewage sludge utilization in agriculture, *Crit. Rev. Environ. Control*, 14, 199, 1984.
80. **Hishida, S.**, Antimony, in *Minor Metals Survey*, Connell, N. J., Ed., Metal Bulletin Ltd., Worcester Park, U.K., 1977, 13.
81. **Henriksson**, Arsenic, in *Minor Metals Survey*, Connell, N. J., Ed., Metal Bulletin Ltd., Worcester Park, U.K., 1977, 19.
82. **Duffus, J. H.**, *Environmental Toxicology*, Edward Arnold, London, 1980, chap. 6.
83. **Clayden, N.**, Bismuth, in *Minor Metals Survey*, Connell, N. J., Ed., Metal Bulletin Ltd., Worcester Park, U.K., 1977, 23.
84. **Paton, R.**, Cadmium, in *Minor Metals Survey*, Connell, N. J., Ed., Metal Bulletin Ltd., Worcester Park, U.K., 1977, 25.
85. **Reichard, H. F. and Spendelow, H. R.**, Chromium, in *Minor Metals Survey*, Connell, N. J., Ed., Metal Bulletin Ltd., Worcester Park, U.K., 1977, 35.
86. **van Grimbergen, F. and Clark, C. P.**, Cobalt, in *Minor Metals Survey*, Connell, N. J., Ed., Metal Bulletin Ltd., Worcester Park, U.K., 1977, 41.
87. **Carter, L. J.**, Manganese, in *Minor Metals Survey*, Connell, N. J., Ed., Metal Bulletin Ltd., Worcester Park, U.K., 1977, 45.
88. **Baltrusaitis, V. A.**, Selenium and tellurium, in *Minor Metals Survey*, Connell, N. J., Ed., Metal Bulletin Ltd., Worcester Park, U.K., 1977, 51.
89. **Zitko, V.**, Toxicity and pollution potential of thallium, *Sci. Total Environ.*, 4, 185, 1975.
90. **Dean, J. G., Bosqui, F. L., and Lanouette, K. H.**, Removing heavy metals from waste water, *Environ. Sci. Technol.*, 6, 518, 1972.
91. Environmental Protection Agency, Inorganic chemicals manufacturing point source category effluent limitations guidelines, pretreatment standards, and new source performance standards, *Fed. Regist.*, 47, 28260, 1982.

92. **Baldi, F. and Borgarali, R.**, Mercury pollution in marine sediments near a chlor-alkali plant: distribution and availability of the metal, *Sci. Total Environ.*, 39, 15, 1984.
93. **Mikac, N., Picer, M., Stegnar, P., and Tusek-Zindarie, Y.**, Mercury distribution in a polluted marine area, ratios of total mercury, methyl mercury and selenium in sediments, mussels and fish, *Water Res.*, 19, 1387, 1985.
94. **Hamza, A., Elsebai, O., and Saleh, H.**, Mercury removal from a chlorolkali plant in Egypt, *Proc. 38th Industrial Waste Conf., Purdue University*, 1983, 339.
95. **Anon.**, Trends in usage of mercury, *J. Metals*, 22, 528, 1970.
96. **Patterson, J. W.**, Effect of carbonate ion on precipitation treatment of cadmium, copper, lead and zinc, *Proc. 36th Waste Conf., Purdue University*, 1981, 579.
97. **Ramirez, E. R. and Koch, F. D.**, Recycling undiluted dragout from toxic plating solutions, *Proc. 37th Ind. Waste Conf., Purdue University*, 1983, 173.
98. **Taylor, C. R. and Qasim, S. R.**, More economical treatment of chromium-bearing wastewaters, *Proc. 37th Ind. Waste Conf., Purdue University*, 1983, 189.
99. **Rabosky, J. G. and Altares, T.**, Wastewater treatment for a small chrome plating shop: a case history, *Proc. 38th Ind. Waste Conf., Purdue University*, 1984, 449.
100. **Sheffield, C. W.**, Treatment of heavy metals at small electroplating plants, *Proc. 36th Ind. Waste Conf., Purdue University*, 1982, 485.
101. **Cartwright, P. S.**, Total effluent treatment and rinse water reclamation in a semiconductor device manufacturing facility, *Water Sci. Technol.*, 17, 325, 1985.
102. **Environmental Protection Agency**, Electroplating and metal finishing point source categories; effluent limitations guidelines, pretreatment standards and new source performance standards, *Fed. Regist.*, 48, 137, 1983.
103. **Vachon, T.**, Removal of iron cyanide from gold mill effluents by ion exchange, *Water Sci. Technol.*, 17, 313, 1985.
104. **Carter, J. L., Jones, M. S., and Tiefenbrunn, R. E.**, Treatment for removal of heavy metals and oil from a piston ring manufacturing waste, *Proc. 37th Ind. Waste Conf., Purdue University*, 1983, 127.
105. **Carlton, J. C., Aronberg, G. R., Washington, D. R., Goodwin, J. K., and Hall, E. P.**, Handling of aluminium forming wastewater, *Proc. 37th Ind. Waste Conf., Purdue University*, 1983, 105.
106. **Cornwell, D. A.**, Alum production from aluminium etching wastes, *Proc. 37th Ind. Waste Conf., Purdue University*, 1983, 119.
107. **Lytle, P. E.**, Fate and speciation of silver in publicly owned treatment works, *Environ. Toxicol. Chem.*, 3, 21, 1984.
108. **Anon.**, *The Silver Market*, Hardy and Harmon, New York, 1982.
109. **Berlow, J. R., Feiler, H. D., and Storch, P. J.**, Paint and ink industry toxic pollutant control, *Proc. 35th Ind. Waste Conf., Purdue University*, 1981, 223.
110. **Shaul, G. M., Barnett, M. W., Neiheisel, T. W., and Dostal, K. A.**, Activated sludge with powdered activated carbon treatment of a dyes and pigments processing wastewater, *Proc. 38th Ind. Waste Conf., Purdue University*, 1984, 659.
111. **Klein, L. A., Lang, M., Nosh, N., and Kirschner, S. L.**, Sources of metals in New York City wastewater, *J. Water Pollut. Control Fed.*, 46, 2653, 1974.
112. **Davis, J. A. and Jacknow, J.**, Heavy metals in wastewater in three urban areas, *J. Water Pollut. Control Fed.*, 47, 2292, 1975.
113. **Yost, K. J. and Wukasch, R. F.**, Pollutant sources and flows in a municipal sewerage system, *Environ. Monit. Assess.*, 3, 61, 1983.
114. **Koch, C. M., Stroka, J. G., Perna, R. K., and Foerster, R. E.**, Impact of pretreatment on sludge content of heavy metals, *J. Water Pollut. Control Fed.*, 54, 339, 1982.
115. **Stonebrook, H., Kerr, J., Newell, G., and Ferell, L.**, Case history of industrial pretreatment in Columbus, Ohio, *J. Water Pollut. Control Fed.*, 56, 1093, 1984.
116. **Brindley, P., Carter, D. C., and Linsmith, L. J.**, A review of industrial effluent control experience, *Water Pollut. Control*, 81, 59, 1982.
117. **Callahan, M. A., Ehreth, D. J., and Levins, P. L.**, Sources of toxic pollutants found in influent to sewage treatment plants, *Proc. 8th Natl. Conf. Municipal Sludge Management*, Information Transfer, Silver Springs, Md., 1979, 55.
118. **Meixmer, G.**, Heavy metals in sewage sludge, *Korrespondenz Abwasser*, 29, 260, 1982.
119. **Mumma, R. O., Raupach, D. C., Waldman, J. P., Tong, S. S. C., Jacobs, M. L., Babisch, J. G., Hotchkiss, J. H., Wszolek, P. C., Gutenamn, W. H., Bache, C. A., and Lisk, D. J.**, National survey of elements and other constituents in municipal sewage sludges, *Arch. Environ. Contamin. Toxicol.*, 13, 75, 1984.
120. **EPA**, Municipal Sludge Management: Environmental Factors, Report MCD-28, EPA 430/9-77-004, Environmental Protection Agency, Washington, D.C., 1977.

121. **Dakers, J. L. and King, R. P.,** Control of toxic metals of sewage treatment works, *Water Sci. Technol.*, 14, 53, 1982.
122. **Harkness, N.,** Problems associated with the disposal of trade effluent to sewers — charging and control, *Water Pollut. Control,* 83, 367, 1984.
123. **Petrasek, A. C. and Kugelman, I. J.,** Metals removals and partitioning in conventional wastewater treatment plants, *J. Water Pollut. Control Fed.,* 55, 1183, 1983.
124. **Patterson, J. W. and Kodukala, P. S.,** Metals distributions in activated sludge systems, *J. Water Pollut. Control Fed.,* 56, 432, 1984.

Chapter 3

POLLUTION CONTROL LEGISLATION

P. W. W. Kirk

TABLE OF CONTENTS

I.	Introduction		66
II.	Water Quality Criteria and Standards		67
	A.	World Health Organization	68
	B.	European Economic Community Directives	69
		1. Council Directive Concerning the Quality of Water Intended for the Abstraction of Drinking Water	70
		2. Council Directive Relating to the Quality of Water Intended for Human Consumption	71
		3. Council Directive on Pollution Caused by Certain Dangerous Substances Discharged into the Aquatic Environment of the Community	71
		4. Council Directive on Limit Values and Quality Objectives for Mercury Discharges by the Chlor-Alkali Electrolysis Industry	73
		5. Council Directive on Limit Values and Quality Objectives for Cadmium Discharges	74
		6. Council Directive on the Protection of Groundwater Against Pollution Caused by Certain Dangerous Substances	74
		7. Environmental Quality Standards Proposed for the U.K.	75
	C.	U.S. Government Criteria and Standards	84
		1. Raw and Potable Waters	84
		2. Wastewaters	88
		3. Other Classes of Water	91
	D.	Water Quality Standards Issued by the Russian Ministry of Health	92
		1. Potable Waters	92
		2. Raw Waters	92
III.	Sludge Disposal to Land		94
IV.	Sludge Disposal to Sea		98
V.	Comparison of Water Quality for Irrigation and Sludge Disposal Guidelines		99
References			101

I. INTRODUCTION

The finite availability of unpolluted freshwater has resulted in considerable pressure for the development of appropriate water quality criteria and standards both to protect potable supply and control the reuse of effluent for a variety of water uses. Heavy metals are ubiquitous in wastewaters and detrimental effects on water quality have been noted in some areas due to the traditional role of surface waters as receiving bodies for wastewater effluents.[1] In addition, surface waters are increasingly being utilized for potable supplies as natural groundwater supplies are depleted and new sources become difficult to exploit. Thus, an increasing trend towards water reuse, both direct and indirect, has resulted.[2,3] Due to the nature of wastewater-treatment processes, accumulation of mineral content is inherent in water reuse;[4,5] thus heavy metals, in particular, give cause for considerable environmental concern due to their nonbiodegradability and toxicity.[6]

The primary objective of the guidelines and legislation reviewed here is to protect public health via the promulgation and enforcement of standards. Protecting raw water quality, as such, represents a change in emphasis which has occurred as a direct result of the increasing indirect reuse of water for potable supply. This change is reflected in the directive promulgated by the Council of the European Communities (CEC) which specifies maximum metal concentrations in raw water intended for potable water abstraction at varying levels of treatment.[7] It is apparent from this directive that limited capability exists in water treatment for the removal of many heavy metals. Indeed the maximum admissible concentrations specified for cadmium, chromium, mercury, lead, and selenium in raw water are identical to those specified for potable water.[8]

There is evidence to suggest that conventional drinking-water purification methods such as sedimentation in clarifying basins, sand filtration, and bank filtration are largely ineffective in reducing heavy-metal concentrations.[1] Reported removals of heavy metals by coagulation-flocculation-decantation-filtration have ranged from only 50% for cobalt, nickel, and lead up to 85% for chromium and copper; although 96% of cadmium was removed.[9] Concern over the efficiency of toxic-metal removal during water treatment has therefore resulted in a realization that metal discharges in municipal wastewater effluent and industrial effluent to surface waters must be controlled in order to safeguard potable supply, particularly in industrialized nations.

The derivation of potable water standards is open to the varying interpretation of available data by each national or international authority. For the most regulated metals, namely arsenic, cadmium, chromium, mercury, lead, and selenium, the World Health Organization (WHO) Guidelines[10] maximum concentrations are identical to those stipulated by the CEC[8], whereas the U.S. Environmental Protection Agency (EPA) standards are less stringent for cadmium and mercury,[11] and the U.S.S.R. standards are more stringent for selenium, but less stringent for lead.[12] The U.S.S.R. standards are, however, applicable to raw waters intended for potable water abstraction, rather than potable supply.

The concept of a "safe" concentration of a toxic contaminant in drinking water is difficult to evaluate, due to the lack of human health data, particularly at low metal concentrations. Carcinogenic risk is generally accepted to have no threshold[13] and therefore any concentration whether it is detectable or undetectable has an associated risk. The EPA has promulgated a list of quality criteria for water which includes estimates of pollutant concentrations which will not affect human health.[14] Although many of these criteria are identical to the existing U.S. drinking-water standards,[11] the criteria concentration for mercury is an order of magnitude below that promulgated for potable water, while the criteria for arsenic and beryllium, based on an incremental increase in cancer risk over a lifetime of 1 in 10^5, are below the detection capabilities of all generally available analytical techniques.[15] Beryllium is not normally regulated in potable water, although the U.S.S.R. has imposed a limit of 0.2 µg/

ℓ of beryllium in potable water (i.e., five times greater than the EPA human health criteria value) for an incremental risk of 1 in 10^5. The criteria for arsenic for an incremental risk of 1 in 10^5 is 3 orders of magnitude in concentration below the potable water limits set by WHO, CEC, EPA, and the U.S.S.R.

Water reuse will inevitably increase the significance of the contribution of drinking water to the total body burden of heavy metals. Environmental sources of exposure to heavy metals include drinking water, diet, and inhalation. Commins[16] estimated the contribution of water in relation to total intake. Based on a water consumption of 3ℓ/day at typical metal concentrations, water may contribute up to 37, 23, 8, 30, and 6% of the total intake of arsenic, cadmium, chromium, lead, and mercury, respectively, while at higher "realistic" metal concentrations, the contribution may rise to 87, 54, 43, 89, and 60%, respectively.

Protection of the aquatic environment has prompted the CEC and EPA to promulgate discharge limits for certain metal-bearing industrial effluents. Discharges of mercury by the chlor-alkali industry for example, are subject to more stringent limits in the U.S. From July 1, 1984 the best available technology economically achievable (BAT) limitations state that the average effluent mercury concentration determined over 30 consecutive days must not exceed 0.1 g/tonne of chlorine production,[17] while the CEC directive stipulates 0.5 g/tonne of production as being the limit of reduction achievable with the current technology.[18] The discharge of cadmium by the pigments industry in the European Economic Community (EEC) will be subject to a limitation of 0.3 mg/ℓ (monthly flow weighted averaged),[19] while a proposed EPA regulation will limit discharges to 0.15 mg/ℓ (30-day average).[20]

The concept of limit values for metals in effluents, although favored by most member states, has not been adopted by the U.K. whose system is based on the application of Environmental Quality Objectives (EQOs).[21] The EQO approach has been reflected in the CEC directives concerning mercury and cadmium discharges which incorporate water-quality objectives applicable to fresh-water, estuarine, and seawater.[18,19] The U.K. Department of the Environment is shortly to issue EQOs for six "gray" list substances — namely, arsenic, chromium, copper, inorganic lead, nickel, and zinc — which specify average concentrations for the protection of freshwater and saltwater aquatic life, irrigation of crops, livestock watering, and abstraction for potable supply or for food processing.[21]

The treatment and disposal of heavy-metal-contaminated sludges is not without attendant risks. Two of the principal disposal sites for sewage sludge are agricultural land and the marine environment. The stability/immobility of heavy metals during sludge treatment and at the ultimate disposal site are critical in determining the safety of the disposal operation. Heavy metals present in sewage sludges give cause for concern in four areas:

1. Human health, when sludge is applied to agricultural land or disposed to sea
2. Adverse effects on biological sludge treatment
3. Adverse effects on the receiving environment (the productivity of agricultural land and the marine environment)
4. Effects on surface and groundwater quality when sludge is applied to land

As a consequence of concern about the above, guidance on the disposal of sewage sludge to land in the U.K. has been given.[22-24] Recently a draft directive has been published by the CEC.[25] Disposal of sewage sludge to the marine environment by the U.K. is regulated by the Dumping at Sea Act,[26] together with the Oslo and London Conventions,[27] while legislation in the U.S. has been introduced to phase out disposal to sea.[28]

II. WATER QUALITY CRITERIA AND STANDARDS

According to the EPA, water-quality criteria "present scientific data and guidance on the environmental effects of pollutants which can be useful to derive regulatory requirements based on considerations of water-quality impacts". In addition, "criteria values do not reflect considerations of economic or technological feasibility".[14] Hence criteria are not regulatory requirements but merely serve as guidelines upon which the regulatory authority may formulate water-quality standards and effluent standards. A water-quality standard is generally accepted as a legally prescribed limit of pollution which has been established under statutory authority. Standards may take into account factors such as social, legal, economic, and hydrological considerations, the environmental and analytical chemistry of the metal, the extrapolation from laboratory data to field situations, and the relationship between the species for which data are available and species which are to be protected.[14]

The legislation relating to concentrations of heavy metals in raw, potable, and wastewaters is diverse and not standardized, owing to the variable nature of the scientific evidence, and interpretation of that evidence, upon which quality criteria and standards have been based. In the following sections, standards and criteria which have been developed by the WHO, the CEC, U.S. government agencies, and the government of the U.S.S.R. are considered. Guidelines and legislation in draft form will be included where appropriate.

A. World Health Organization

The European[29] and international[30] WHO "standards" for potable water have now been replaced by the recently published *Guidelines for Drinking Water Quality*.[10] These guidelines recommended heavy-metal concentrations which should not be exceeded in potable waters. It is emphasized in the guidelines that the values quoted will not be mandatory but will aim to set minimum requirements for the quality of water intended for domestic use, upon which national legislation may be based. It is recognized in the guidelines that certain areas may require time to attain the stated values and that inferior interim standards may be adopted. However, the ultimate aim would be to attain at least the recommended values.[31]

In developing the guideline values for metals, the objective was to define a quality of water which could safely be consumed over a lifetime of exposure. Specific factors considered when setting such a guideline value included:

1. Scientific criteria, defining dose-response relationships
2. Data concerning frequency of occurrence and concentrations of contaminants commonly found in drinking water
3. Knowledge of the feasibility of applying control techniques to remove or reduce the concentration in drinking water

It is acknowledged that the guidelines are open to a degree of uncertainty mainly as a result of:

1. Validity of the scientific data
2. Extrapolation of dose-response relationships to low levels not experimentally verified
3. Extrapolation of dose-response relationships from animal data to man
4. Uncertainty concerning the division of intake between air, water, food, and other sources (e.g., occupational)

In the guidelines, maximum permissible concentrations are specified for arsenic, cadmium, chromium, lead, mercury, and selenium based on toxicological data; these values are compared with both the superceded international and European "standards" in Table 1. Several of the guideline values are more stringent than the WHO "standards" which they replace. The guideline value for lead, for example, has been halved since an exposure of 0.1 mg/ℓ of lead in drinking water may result in blood lead concentrations in a significant number of

Table 1
METALS INCLUDED IN THE WHO INTERNATIONAL AND EUROPEAN "STANDARDS" AND GUIDELINES FOR DRINKING-WATER QUALITY[10,29,30]

Metal	International HD (mg/ℓ)	International MP (mg/ℓ)	European (MP [mg/ℓ])	Guidelines (MP [mg/ℓ])
As	—	0.05	0.05	0.05
Cd	—	0.01	0.01	0.005
Cr	—	—	0.05[a]	0.05[b]
Hg	—	0.001	PC	0.001
Pb	—	0.1	0.1	0.05
Se	—	0.01	0.01	0.01
Ag	—	—	PC	—
Ba	—	PC	1.0	—
Be	—	PC	PC	—
Co	—	PC	—	—
Mo	—	PC	PC	—
Sn	—	PC	PC	—
U	—	PC	PC	—
V	—	PC	PC	—
Al[c]	—	—	—	0.2
Cu[c]	0.05	1.5	0.05	1.0
Fe[c]	0.1	1.0	0.1	0.3
Mn[c]	0.05	0.5	0.05	0.1
Zn[c]	5.0	15	5.0	5.0

Note: HD, highest desirable concentration; MP, maximum permissible concentration; PC, presence should be controlled.

[a] Cr(VI).
[b] Total Cr.
[c] Values based on aesthetics, corrosion, deposit formation, etc.

children which exceed the recommended level of lead in the blood of 0.03 mg/100 mℓ. It has been estimated that lead contributions from drinking water may constitute one third of ingested and absorbed lead in children, although the relationship between lead in blood and defined exposures to drinking water is not well documented. Chromium, on the other hand, is now stated in terms of total metal in the guidelines as opposed to Cr(VI), as previously expressed in the WHO "standards", due to analytical difficulties in determining the more toxic Cr(VI) form.

More subjective guidelines have been proposed with regard to metals causing aesthetic problems. Healthy individuals appear to be unaffected by orally ingested aluminum compounds, although neurological disorders in renal-dialysis patients have been noted and discoloration of water may occur above 0.1 mg/ℓ. Copper, an essential element, can corrode galvanized iron and steel fittings while iron precipitates color the water and leave deposits. Manganese causes staining above 0.15 mg/ℓ and zinc imparts an astringent taste above 5.0 mg/ℓ. Guideline values for five metals based on aesthetic considerations are compared with the old WHO "standards" concentrations for these metals in Table 1.

B. European Economic Community Directives

Member states of the EEC are obliged to modify or adapt their national legislation in order to comply with directives from the CEC within stated time limits. Under community law, the CEC can make decisions (binding in their entirety) or recommendations (binds the

Table 2
CEC QUALITY REQUIREMENTS OF SURFACE WATERS FOR DRINKING WATER ABSTRACTION[7]

Metal (mg/ℓ)	A1 GL	A1 MAC	A2 GL	A2 MAC	A3 GL	A3 MAC
As	0.01	0.05	—	0.05	0.05	0.1
Ba	—	0.1	—	1	—	1
B	1	—	1	—	1	—
Cd	0.001	0.005	0.001	0.005	0.001	0.005
Cr	—	0.05	—	0.05	—	0.05
Cu	0.02	0.05	0.05	—	1	—
Fe	0.1	0.3	1	2	1	—
Hg	0.0005	0.001	0.0005	0.001	0.0005	0.001
Mn	0.05	—	0.1	—	1	—
Pb	—	0.05	—	0.05	—	0.05
Se	—	0.01	—	0.01	—	0.01
Zn	0.5	3	1	5	1	5

Note: Definition of the standard methods of treatment for transforming surface water of categories A1, A2, and A3 into drinking water:
A1 simple physical treatment and disinfection (e.g., rapid filtration and chlorination)
A2 normal physical treatment, chemical treatment, and disinfection (e.g., prechlorination, coagulation, flocculation, decantation, filtration, disinfection [final chlorination])
A3 intensive physical and chemical treatment, extended treatment and disinfection, (e.g., chlorination to breakpoint, coagulation, flocculation, decantation, filtration, adsorption [activated carbon], disinfection [ozone, final chlorination]).
GL = guide level. MAC = maximum admissible concentration.

end result but not the method of achieving), or issue opinions (not binding). Directives, under the Rome Treaty, are binding to the member states regarding the results that have to be achieved, but leave the form and methods of achieving it to national legislation. Member states must set values applicable to the various types of water for parameters listed in the directives which are less than or equal to the stipulated maximum admissible concentrations. It is, however, recommended that standards in individual member states should be set with regard to the lower guide levels.

1. Council Directive Concerning the Quality of Water Intended for the Abstraction of Drinking Water

This directive relates to the quality of surface water intended for the abstraction of potable supplies in the member states.[7] Under the articles of this directive, which came into force in June 1977, each member state must apply the mandatory values specified without any distinction to national waters and waters crossing its frontiers. Groundwater, brackish water, and water intended to recharge aquifiers are, however, specifically excluded.

Surface waters are divided according to the extent of treatment they require for the production of potable supplies. The metals included in the three categories A1, A2, and A3 are listed in Table 2, which specifies mandatory and guide values. The intention of the directive is to achieve an improvement in surface-water quality and as such requires the submission of national plans for surface-water improvement, with priority being given to class A3 waters. Under Article 5, sampling values must demonstrate that the water complies with the parametric values for the water quality parameters in the case of 95% of samples for parameters conforming to those specified in the "MAC" columns of Table 2 and 90% of the samples in all other cases. In the case of the 5 to 10% of samples which do not comply (1) the value must not deviate from the parametric value by more than 50%, (2) there can be no resultant danger to public health, and (3) consecutive water samples should not deviate from the relevant parametric value.

Frequency of sampling and analytical methodology required to enforce the directive will be the subject of a further publication by the commission. Until such time, they will be defined by competent national authorities.

2. Council Directive Relating to the Quality of Water Intended for Human Consumption

This publication is intended to safeguard potable supplies and water used in the food production and processing industries but excluded natural mineral waters and medicinal water.[8] A total of 16 metals are included on the grounds of direct toxicity or undesirability. Guide levels and maximum admissible concentrations are given in Table 3. According to Article 7 of the directive, the values fixed by the member states must be less than or the same as the values shown in the maximum admissible concentration column, while the guide level must also be taken into consideration. Member states must have ensured that all necessary measures were taken to comply with this directive within 5 years of its notification (i.e., by July 1985), according to Article 19, and each state must have undertaken regular monitoring of potable supplies.

3. Council Directive on Pollution Caused by Certain Dangerous Substances Discharged into the Aquatic Environment of the Community

The aim of this directive is to protect inland surface waters, territorial waters, and internal coastal waters from the effects of discharges of specific toxic substances.[32] Provisions relating to groundwater have now been included in a separate directive on groundwater (80/68/EEC).[33] The directive introduced a "black" (I) list which was largely compiled on the basis of toxicity, persistence, and bioaccumulation and a "gray" (II) list of substances known to have a deleterious effect on the aquatic environment but whose effects could be confined to a limited area. Member states are required to take appropriate steps to eliminate pollution of these waters by List I substances and to reduce pollution by the substances in List II, with the povisions of this directive representing only a first step towards this goal. Metals and metal compounds included in Lists I and II are presented in Tables 4 and 5.

All discharges containing List I substances require authorization by the competent authority of the member state, who must apply limit values which may not exceed those specified by the council. In the case of existing discharges time limits will be given for the implementation of these values. The limit values will be expressed in terms of maximum concentrations of List I substances in the discharges and, where appropriate, the maximum quantity of each substance expressed as a unit weight of the pollutant per unit weight of the characteristic element of polluting activity (e.g., unit weight per unit of raw material or per product unit). Limit values will mainly be set on the basis of toxicity, persistence, and bioaccumulation, but will take into account the best technical means available.

Water-quality objectives for List I substances will also be laid down, and indeed if these objectives are being met and continuously maintained throughout the area in question, the limit values discussed previously may not be applied.

In order to reduce pollution by List II substances, member states are required to adhere to emission standards based on water-quality objectives, in accordance with council direc-

Table 3
METALS LISTED IN THE CEC DIRECTIVE RELATING TO WATER QUALITY FOR HUMAN CONSUMPTION[8]

Metal	Guide level (mg/ℓ)	MAC (mg/ℓ)	Comments
As[a]		0.05	
Cd[a]		0.005	
Cr[a]		0.05	
Hg[a]		0.001	
Ni[a]		0.05	
Pb[a]		0.05	Value after flushing (if direct or after flushing frequently or significantly exceeds 0.1 mg/ℓ, action must be taken by consumer)
Sb[a]		0.01	
Se[a]		0.01	
Ag[b]		0.01	If exceptionally used non-systematically to process water, MAC value of 0.08 mg/ℓ may be authorized
Al[b]	0.05	0.2	
Ba[b]	0.1		
B[b]	1		
Cu[b]	0.1[c] 3[d]		Above 3 mg/ℓ, astringent taste, discoloration and corrosion may occur
Fe[b]	0.05	0.2	
Mn[b]	0.02	0.05	
Zn[b]	0.1[c] 5[d]		Above 5 mg/ℓ astringent taste, opalescence and sand-like deposits may occur

Note: MAC, maximum admissible concentration.

[a] Values based on toxicity.
[b] Values based on other parameter.
[c] At outlets of pumping and/or treatment works and their substations.
[d] After the water has been standing for 12 hr in the piping and at the point where the water is made available to the consumer.

tives, where they exist. Programs incorporating these objectives and which take into account the latest economically feasible technical developments must be submitted to the council for approval, and deadlines set for their implementation. Emission standards are to be laid down for discharges of substances in both lists into sewers, where this is necessary for the implementation of this directive.

Subsequent to this directive, a series of directives will deal with specific List I substances (those for mercury and cadmium having already been produced).[18,19] The council, acting on a proposal from the commission or at the request of a member state, will revise and, where necessary, supplement List I and II. If appropriate, substances may be transferred from List II to List I.

Table 4
METALS AND METAL COMPOUNDS INCLUDED IN THE CEC "BLACK" LIST[32]

1. Organotin compounds
2. Substances whose carcinogenic activity is exhibited in or by the intervention of the aquatic environment[a]
3. Mercury and its compounds
4. Cadmium and its compounds

[a] Where certain substances in List II are carcinogenic, they are included in Category 2 of this list.

Table 5
METALS AND METAL COMPOUNDS INCLUDED IN THE CEC "GRAY" LIST[32]

A. The following metalloids and metals and their compounds:

1. Zn	6. Se	11. Sn	16. V
2. Cu	7. As	12. Ba	17. Co
3. Ni	8. Sb	13. Be	18. Tl
4. Cr	9. Mo	14. B	19. Te
5. Pb	10. Ti	15. U	20. Ag

B. Substances which have a deleterious effect on the taste and/or smell of products for human consumption derived from the aquatic environment and compounds liable to give rise to such substances in water.

Table 6
LIMIT VALUES AND TIME LIMITS FOR COMPLIANCE REQUIRED BY THE CEC DIRECTIVE ON MERCURY DISCHARGES BY THE CHLOR-ALKALI ELECTROLYSIS INDUSTRY[18]

Parameter	Monthly[a] average limit values not to be exceeded		Comments
	July 1, 1983	July 1, 1986	
Limit values for recycled and lost brine in terms of concentration which, in principle, should not be exceeded (mg/ℓ)	0.075	0.05	In all cases these limit values may not be greater than those expressed as maximum quantities divided by water required per tonne of installed chlorine production capacity
Limit values for recycled brine in terms of grams of mercury discharged per tonne of installed chlorine production capacity	0.5	0.5	Applicable to effluent from chlorine production unit
		1.0	Applicable to total discharge from site
Limit values for lost brine in terms of grams of mercury per tonne of installed chlorine production capacity	8.0	5.0	Applicable to total discharge from site

[a] Daily average limit values are also enforced which correspond to four times the appropriate monthly average limit values.

4. Council Directive on Limit Values and Quality Objectives for Mercury Discharges by the Chlor-Alkali Electrolysis Industry

This directive is specifically aimed at the chlor-alkali electrolysis industry, which is a major source of mercury discharged to the aquatic environment.[18] Member states may limit mercury discharges by adopting effluent-quality standards (Table 6) or by achieving appropriate water-quality objectives according to Article 6 of Directive 76/464/EEC.[32] Applying the latter approach, a competent authority will determine the area affected by each discharge and select one or more of the following quality objectives, having regard to the intended use of the affected area and the aims of the directive to limit pollution of the aquatic environment by mercury.

1. The concentration of mercury in a representative sample of fish flesh must not exceed 0.3 mg/kg wet flesh
2. Total mercury concentration in inland surface waters must not exceed 1 µg/ℓ
3. Soluble mercury in estuarine waters must not exceed 0.5 µg/ℓ
4. Soluble mercury in territorial sea waters and internal coastal waters must not exceed 0.3 µg/ℓ
5. Water quality must comply with any other council directives applicable to such waters as regards the presence of mercury

The values in objectives 2, 3, and 4, which are arithmetic means over 1 year, could as an exception and where necessary for technical reasons, be multiplied by 1.5 until June 30, 1986. Where several objectives are applied, water quality must be sufficient to meet each of them and in all cases the concentration of mercury in sediments or in shellfish must not increase significantly with time. Member states must have complied with the Articles of this Directive by July 1, 1983.

5. Council Directive on Limit Values and Quality Objectives for Cadmium Discharges

This directive applies to discharges of cadmium both in terms of effluent concentration and total cadmium discharge per unit of cadmium handled.[19] Member states must have enforced the articles of this directive by September 26, 1985. It applies to inland surface water, estuarine, territorial, and internal coastal waters but specifically excludes groundwater, which is the subject of the separate Directive 80/68/EEC.[33] In accordance with Directive 76/464/EEC,[32] member states may apply effluent standards as presented in Table 7 or alternatively adopt one or more water-quality objectives specified in Table 8. In addition, waters used for the abstraction of drinking water must conform to the requirements of Directive 75/440/EEC[7] and the concentration of cadmium in sediments and/or shellfish (if possible *Mytilus edulis*) should not increase with time, where water-quality objectives are adopted. When several quality objectives are applied, water quality must be sufficient to comply with each of the objectives.

6. Council Directive on the Protection of Groundwater Against Pollution Caused by Certain Dangerous Substances

This directive, which came into force in December 1981, applies only to groundwater which is defined as "water which is below the surface of the ground in the saturation zone and in direct contact with the ground or subsoil".[33] It specifically excludes, however, discharges of domestic effluents from isolated dwellings situated outside areas protected for the abstraction of potable water and discharges found by a competent authority of the member state not to contain concentrations of List I and List II substances which could result in any present or future danger of deterioration in the quality of the receiving groundwater.

Member states must prevent contamination of groundwaters by List I substances by prohibiting all direct discharges (without percolation through the ground or subsoil) of substances in List I and investigate any disposal operation which might lead to indirect discharges (following percolation through the ground or subsoil), except where the groundwater into which the discharge is taking place is permanently unsuitable for other uses. Any discharge of List II substances directly or indirectly to groundwaters is also subject to prior investigation which may include examination of the hydrogeological conditions of the area concerned, together with the potential purifying capacity of the soil and subsoil and the risk of pollution and alteration of groundwater quality.

Discharge or disposal authorizations, which must be reviewed at least every 4 years and granted only for limited periods, are subject to a number of specifications. These include the place and method of discharge or disposal, essential precautions required, maximum

Table 7
LIMIT VALUES AND TIME LIMITS FOR COMPLIANCE REQUIRED BY THE CEC DIRECTIVE ON CADMIUM DISCHARGES[19]

Industrial sector[a]	Units	Limit values must be complied with as from 1.1.1986	1.1.1989[b]
Zinc mining, lead and zinc refining, cadmium metal, and nonferrous metal industry	mg/ℓ	0.3[c]	0.2[c]
Manufacture of cadmium compounds and stabilizers	mg/ℓ	0.5[c]	0.2[c]
	g/kg	0.5[d]	—[e]
Manufacture of pigments, electroplating[f]	mg/ℓ	0.5[c]	0.2[c]
	g/kg	0.3[d]	—[e]
Manufacture of primary and secondary batteries	mg/ℓ	0.5[c]	0.2[c]
	g/kg	1.5[d]	—[e]
Manufacture of phosphoric acid and/or phosphoric fertilizer from phosphatic rock[g]	—	—	—

[a] Limit values for other industrial sectors may be fixed by the council. Meanwhile member states should fix emission standards autonomously in accordance with Directive 76/464/EEC.[32] Such standards must take into account the best technical means available and must not be less stringent than the most nearly comparable limit value here.
[b] More restrictive limit values may come into force in 1992.
[c] Monthly flow-weighted average concentration of total cadmium in discharge.
[d] Monthly average grams of cadmium discharged per kilogram of cadmium handled.
[e] Limit values expressed as loads will remain at the 1.1.1986 level unless otherwise specified.
[f] Member states may suspend application of limit values for electroplating until 1.1.1989 for plants which discharge less than 10 kg of cadmium per year and in which the total volume of the electroplating tanks is less than 1.5 m^3, if technical or administrative considerations make such a step absolutely necessary.
[g] There is presently no economically feasible technical method of extracting cadmium from these wastes. Member states should, however, fix emission standards for these discharges in accordance with Directive 76/464/EEC.[32]

quantities and concentrations of controlled substances, and monitoring arrangements for effluents or groundwater quality as appropriate. In the case of tipping or disposal, technical precautions required to prevent any discharge of List I substances and any pollution as a result of discharge of List II substances to groundwater must be specified.

7. Environmental Quality Standards Proposed for the U.K.

In response to the requirement set out in the CEC Dangerous Substances Directive[32] the U.K. Water Research Centre has issued a series of technical reports containing recommendations for Environmental Quality Standards (EQSs) appropriate for specific water uses. A technical report has been issued for each of the List II metals arsenic,[34] chromium,[35] copper,[36] inorganic lead,[37] nickel,[38] and zinc.[39] Tables 9 to 14 summarize these recommendations, although the original source documents should be referred to for a complete description.

Table 8
WATER QUALITY OBJECTIVES REQUIRED BY THE CEC DIRECTIVE ON CADMIUM DISCHARGES TO THE AQUATIC ENVIRONMENT[19]

Water category	Cadmium form	Cadmium concentration[a] (mg/ℓ)	
		Waters affected by discharges	Waters in national monitoring network
Inland surface	Total	0.005	0.001
Estuarine	Dissolved	0.005	0.001
Territorial and internal coastal	Dissolved	0.0025	0.005

[a] Arithmetic mean of results obtained over 1 year.

Table 9
EQS VALUES RECOMMENDED BY THE U.K. WATER RESEARCH CENTRE FOR ARSENIC

Use	EQS value (μg As/ℓ)			Dissolved (D) or total (T)
	Av Conc	95-percentile conc	Max allowable conc	
Freshwater				
Direct abstraction to potable supply				
Treatment Category[a]				
A1	—	50*	75*	T
A2	—	50*	75*	T
A3	—	100*	150*	T
Abstraction to a reservoir for potable supply[b]	—	$\frac{50 \times 100}{100 - R}$	—	T
Food for human consumption derived from freshwater[c]	—	—	—	—
Protection of freshwater fish[d]	50	—	—	D
Protection of other freshwater life and associated nonaquatic organisms				
Invertebrates	50	—	—	D
Plants	150	—	—	D
Irrigation of crops[e]	40	—	—	T
Livestock watering[f]	200	—	—	T
Abstraction for food-processing industry[g]	—	50	—	T
Bathing and contact water sports[h]	—	500	—	T
Saltwater				
Food for human consumption derived from saltwater[c]	—	—	—	—
Protection of saltwater fish and shellfish[i]	25	—	—	D
Protection of other saltwater life and associated non-aquatic organisms[i]	25	—	—	D
Bathing and contact water sports[h]	—	500	—	T

Note: * indicates a mandatory value. EQS, environmental quality standard.

[a] Required by the directive concerning the quality of surface water intended for the abstraction of drinking water in the member states.[7] Derogations allowed for floods, other natural disasters, and natural enrichment.

Table 9 (continued)
EQS VALUES RECOMMENDED BY THE U.K. WATER RESEARCH CENTRE FOR ARSENIC

[b] R is the average extent of removal of arsenic (percent) during passage through the reservoir. Water abstracted from the reservoir must comply with the Surface Water Directive.[7]
[c] The maximum estimated intake from fish and shellfish is not thought to present a risk to human health.
[d] Other values may be appropriate to local situations, e.g., for the protection of particularly sensitive species. Arsenic is not one of the parameters covered by the Directive of 18th July 1978 on the quality of freshwaters needing protection or improvement in order to support fish life.[40]
[e] ADAS leaflet No. 776.[41] The value given is the suggested maximum average concentration during the normal irrigation period (i.e., the summer months). If exceeded, soil levels should be monitored.
[f] In order to make the standards for livestock watering consistent with others in Table 9, the "desirable" level has been interpreted as the annual average concentration. An "acceptable" level would be 500 µg/ℓ and a "maximum" tolerated level of 5 mg/ℓ has been suggested.
[g] Other levels, to be derived locally, may be appropriate for individual industries and plants.
[h] This standard is based on the assumption that no more than one tenth of the water consumed by any individual is likely to be water accidentally swallowed during bathing or contact water sports. For most individuals, the consumption of water (and therefore arsenic) by this mechanism would be negligible.
[i] Other values may be appropriate to local situations, e.g., for the protection of particularly sensitive species or shellfisheries.

Modified from Mance, G., Musselwhite, C., and Brown, V. M., *Proposed Environmental Quality Standards for List II Substances in Water: Arsenic*, TR212, Water Research Centre, Medmenham, U.K., 1984. With permission.

Table 10
EQS VALUES RECOMMENDED BY THE U.K. WATER RESEARCH CENTRE FOR TOTAL CHROMIUM (III) AND (VI)

Use	Av conc	95-percentile conc	Max allowable conc	Dissolved (D) or Total (T)
Freshwater				
Direct abstraction to potable supply[a]	—	50*	75*	T
Abstraction to a reservoir for potable supply[b]	—	$\frac{50 \times 100}{100 - R}$	—	T
Food for human consumption derived from freshwater[c]	—	—	—	—
Protection of freshwater fish[d]				
Salmonid fish				
Total hardness (mg CaCO$_3$/ℓ)				
<50	5	—	—	D
50—100	10	—	—	D
100—200	20	—	—	D
>200	50	—	—	D
Coarse fish				
Total hardness (mg CaCO$_3$/ℓ)				
<50	150	—	—	D
50—100	175	—	—	D
100—200	200	—	—	D
>200	250	—	—	D

Table 10 (continued)
EQS VALUES RECOMMENDED BY THE U.K. WATER RESEARCH CENTRE FOR TOTAL CHROMIUM (III) AND (VI)

	EQS value (μg Cr/ℓ except where stated)			
Use	Av conc	95-percentile conc	Max allowable conc	Dissolved (D) or Total (T)
Protection of other freshwater life and associated nonaquatic organisms Total hardness (mg CaCO$_3$/ℓ)				
<50	5	—	—	D
50—100	10	—	—	D
100—200	20	—	—	D
>200	50	—	—	D
Irrigation of crops[e]	2000	—	—	T
Livestock watering[f]	1000	—	—	T
Abstraction for food-processing industry[g]	—	50	—	T
Bathing and contact water sports[h]	—	500	—	T
Saltwater				
Protection of saltwater fish and shellfish[i]	15	—	—	D
Protection of other saltwater life and associated nonaquatic organisms[i]	15	—	—	D
Bathing and contact water sports[h]	—	500	—	T
Food for human consumption derived from saltwater[c]	—	—	—	—

Note: * indicates a mandatory value. EQS, environmental quality standard.

[a] Required by the directive concerning the quality of surface water intended for the abstraction of drinking water in the member states.[7] Derogations allowed for floods, other natural disasters, and natural enrichment.

[b] R is the average extent of removal of chromium (percent) during passage through the reservoir. Water abstracted from the reservoir must comply with the Surface Water Directive.[7]

[c] The maximum estimated intake from fish and shellfish is not thought to present a risk to human health.

[d] Other values may be appropriate to local situations, e.g., for the protection of particularly sensitive species. Chromium is not one of the parameters covered by the Directive of 18th July 1978 on the quality of freshwaters needing protection or improvement in order to support fish life.[40]

[e] ADAS leaflet No. 776.[41] The value given is the suggested maximum average concentration during the normal irrigation period (i.e., the summer months). If exceeded, soil levels should be monitored.

[f] In order to make the standards for livestock watering consistent with others in Table 10, the "desirable" level has been interpreted as the annual average concentration. A "maximum" tolerated level of 5000 mg/ℓ has been suggested for most stock.

[g] Other levels, to be derived locally, may be appropriate for individual industries and plants.

[h] This standard is based on the assumption that no more than one tenth of the water consumed by any individual is likely to be water accidentally swallowed during bathing or contact water sports. For most individuals, the consumption of water (and therefore chromium) by this mechanism would be negligible.

[i] Other values may be appropriate to local situations, e.g., for the protection of particularly sensitive species or shellfisheries.

Modified from Mance, G., Brown, V. M., Gardiner, J., and Yates, J., *Proposed Environmental Quality Standards for List II Substances in Water: Chromium*, TR207, Water Research Centre, Medmenham, U.K., 1984. With permission.

Table 11
EQS VALUES RECOMMENDED BY THE U.K. WATER RESEARCH CENTRE FOR TOTAL COPPER

Use	EQS value (μg Cu/ℓ except where stated)			
	Av conc	95-percentile conc	Max allowable conc	Dissolved (D) or total (T)
Freshwater				
Direct abstraction to potable supply				
Treatment Category[a]				
A1	—	50	75	T
A2 (guide values)	(50)	—	—	T
A3 (guide values)	(1000)	—	—	T
Abstraction to a reservoir for potable supply[b]	—	$\frac{50 \times 100}{100 - R}$	—	T
Food for human consumption derived from freshwater[c]	—	—	—	—
Protection of freshwater fish[d]				
Total hardness (mg CaCO$_3$/ℓ)				
10	1	5*	—	D
50	6	22*	—	D
100	10	40*	—	D
300	28	112*	—	D
Protection of other freshwater life and associated nonaquatic organisms[e]				
Total hardness (mg CaCO$_3$/ℓ)				
10	1	—	—	D
50	6	—	—	D
100	10	—	—	D
300	28	—	—	D
Irrigation of crops[f]	500	—	—	T
Livestock watering[g]	200	—	—	T
Abstraction for food processing industry[h]	—	50	—	T
Bathing and contact water sports[i]	—	500	—	T
Saltwater				
Protection of saltwater fish and shellfish[j]	5	—	—	D
Protection of other saltwater life and[j] associated nonaquatic organisms	5	—	—	D
Bathing and contact water sports[i]	—	500	—	T
Food for human consumption derived from saltwater[c]	—	—	—	—

Note: * indicates a mandatory value. EQS, environmental quality standard.

[a] Required by the directive concerning the quality of surface water intended for the abstraction of drinking water in the member states.[7] Derogations allowed for floods, other natural disasters, and natural enrichment.
[b] R is the average extent of removal of copper (percent) during passage through the reservoir. Water abstracted from the reservoir must comply with the Surface Water Directive.[7]
[c] The maximum estimated intake from fish and shellfish is not thought to present a risk to human health.
[d] Copper is one of the determinands covered by the Directive of 18th July 1978 on the quality of freshwaters needing protection or improvement in order to support fish life.[40] Derogation is allowed for natural enrichment. It is stressed that the acceptable concentration may be significantly higher in the presence of organic matter. Mandatory values apply to designated fisheries only.
[e] The acceptable concentration may be significantly higher in the presence of organic matter.
[f] ADAS leaflet No. 776.[41] The value given is the suggested maximum average concentration during the normal irrigation period (i.e., the summer months). If exceeded, soil levels should be monitored.

Table 11 (continued)
EQS VALUES RECOMMENDED BY THE U.K. WATER RESEARCH CENTRE FOR TOTAL COPPER

g In order to make the standards for livestock watering consistent with others in Table 11, the "desirable" level has been interpreted as the annual average concentration. An "acceptable" level might be as high as 10 mg/ℓ for short periods of time and a "maximum tolerated level" of 25 mg/ℓ has been suggested.

h Other levels, to be derived locally, may be appropriate for individual industries and plants.

i This standard is based on the assumption that no more than one tenth of the water consumed by any individual is likely to be water accidentally swallowed during bathing or contact water sports. For most individuals, the consumption of water (and therefore copper) by this mechanism would be negligible.

j Other values may be appropriate to local situations, e.g., for the protection of particularly sensitive species or shellfisheries. Higher concentrations are acceptable in areas where a history of copper contamination has allowed acclimitization, or where the presence or organic matter may lead to complexation of the copper.

Modified from Mance, G., Brown, V. M., and Yates, J., *Proposed Environmental Water Quality Standards for List II Substances in Water: Copper*, TR210, Water Research Centre, Medmenham, U.K., 1984. With permission.

Table 12
EQS VALUES RECOMMENDED BY THE U.K. WATER RESEARCH CENTRE FOR INORGANIC LEAD

	EQS value (μg Pb/ℓ except where stated)			
Use	Av conc	95-percentile conc	Max allowable conc	Dissolved (D) or total (T)
Freshwater				
Direct abstraction to potable supply[a]	—	50*	75	T
Abstraction to a reservoir for potable supply[b]	—	$\frac{50 \times 100}{100 - R}$	—	T
Food for human consumption derived from freshwater[c]				
Fresh fish	—	—	2 mg/kg*	—
Shellfish	—	—	10 mg/kg*	—
Game, game pâté	—	—	10 mg/kg*	—
Watercress, seaweeds, and other vegetable products	—	—	1 mg/kg*	—
Protection of freshwater fish[d]				
Salmonid fish				
Total hardness (mg CaCO$_3$/ℓ)				
<50	4	—	—	D
50—150	10	—	—	D
>150	20	—	—	D
Coarse fish				
Total hardness (mg CaCO$_3$/ℓ)				
<50	50	—	—	D
50—150	125	—	—	D
>150	250	—	—	D
Protection of other freshwater life and associated nonaquatic organisms				
Total hardness (mg CaCO$_3$/ℓ)				
<75	5	—	—	D
>75	60	—	—	D
Irrigation of crops[e]	2000	—	—	T
Livestock watering[f]	100	—	—	T

Table 12 (continued)
EQS VALUES RECOMMENDED BY THE U.K. WATER RESEARCH CENTRE FOR INORGANIC LEAD

	EQS value (μg Pb/ℓ except where stated)			
Use	Av conc	95-percentile conc	Max allowable conc	Dissolved (D) or total (T)
Abstraction for food processing industry[g]	—	50	—	T
Bathing and contact water sports[h]	—	500	—	T
Saltwater				
Protection of saltwater fish and shellfish[i]	25	—	—	D
Protection of other saltwater life and associated nonaquatic organisms[i]	25	—	—	D
Bathing and contact water sports[h]	—	500	—	T
Food for human consumption derived from saltwater[c]				
Fresh fish	—	—	2 mg/kg*	—
Shellfish	—	—	10 mg/kg*	—
Game, game pâté	—	—	10 mg/kg*	—
Watercress, seaweeds, and other vegetable products	—	—	1 mg/kg*	—

Note: * indicates a mandatory value. EQS, environmental quality standard. The EQS values given for various uses of water assume that the lead is present almost entirely as inorganic lead. If a significant proportion of organolead is present, different EQS values (not given here) would be required.

[a] Required by the directive concerning the quality of surface water intended for the abstraction of drinking water in the member states.[7] The same EQS value applies, irrespective of water treatment process employed. Derogations allowed for floods, other natural disasters and natural enrichment.

[b] R is the average extent of removal of total lead (percent) during passage through the reservoir. Water abstracted from the reservoir must comply with the Surface Water Directive.[7]

[c] Maximum lead contents of many varieties of food are specified in The Lead in Food Regulations, which came into operation on April 12, 1980.[42] For those categories of food not specifically named, a maximum allowable concentration of 1 mg/kg applies. Responsibility for monitoring rests with the Ministry of Agriculture, Fisheries, and Foods, the Department of Agriculture and Fisheries for Scotland, and the public health authorities.

[d] Other values may be appropriate to local situations, e.g., for the protection of particularly sensitive species or when organolead is present as a significant proportion of the total dissolved lead. Lead is not one of the parameters covered by the Directive of 18th July 1978 on the quality of freshwaters needing protection or improvement in order to support fish life.[40] Where breeding populations of rainbow trout are present, the EQS should be 50% of that recommended for normal salmonid waters.

[e] ADAS leaflet No. 776.[41] The value given is the suggested maximum average concentration during the normal irrigation period (i.e., the summer months). If exceeded, soil levels should be monitored.

[f] In order to make the standard for livestock watering consistent with others in Table 12, the "desirable" level has been interpreted as the annual average concentration. An "acceptable" level of 500 μg/ℓ Pb has been suggested with a "maximum tolerated level" (MTL) of 30 mg/ℓ Pb. This MTL applies to most stock, however for sheep, the MTL is 3 mg/ℓ Pb.

[g] Other levels, to be derived locally, may be appropriate for individual industries and plants.

[h] This standard is based on the assumption that no more than one tenth of the water consumed by any individual is likely to be water accidentally swallowed during bathing or contact water sports. For most individuals the consumption of water (and therefore lead) by this mechanism would be negligible.

[i] Other values may be appropriate to local situations, e.g., for the protection of particularly sensitive species or when organolead is present as a significant proportion of the total dissolved lead.

Modified from Brown, V. M., Gardiner, J., and Yates, J., *Proposed Environmental Quality Standards for List II Substances in Water: Inorganic Lead*, TR208, Water Research Centre, Medmenham, U.K., 1984. With permission.

Table 13
EQS VALUES RECOMMENDED BY THE U.K. WATER RESEARCH CENTRE FOR NICKEL

	EQS value (μg Ni/ℓ except where stated)			
Use	Average conc	95-percentile conc	Max allowable conc	Dissolved (D) or total (T)
Freshwater				
Direct abstraction to potable supply[a]	(50)	—	—	T
Protection of freshwater fish[b]				
Total hardness (mg CaCO$_3$/ℓ)				
<50	50	—	—	D
50—100	100	—	—	D
100—200	150	—	—	D
>200	200	—	—	D
Food for human consumption derived from freshwater[c]	—	—	—	—
Protection of other freshwater life and associated non-aquatic organisms[d]				
Total hardness (mg CaCO$_3$/ℓ)				
<50	8	—	—	D
50—100	20	—	—	D
100—200	50	—	—	D
>200	100	—	—	D
Irrigation of crops[e]	150	—	—	T
Livestock watering[f]	1000	—	—	T
Abstraction for food processing industry[g]	(50)	—	—	T
Bathing and contact water sports[h]	500	—	—	T
Saltwater				
Protection of saltwater fish and shellfish[i]	30	—	—	D
Protection of other saltwater life and associated nonaquatic organisms[i]	30	—	—	D
Bathing and contact water sports[h]	500	—	—	T
Food for human consumption derived from saltwater[c]	—	—	—	—

Note: EQS, environmental quality standard.

[a] The standard applies to the water quality AFTER treatment as in the Drinking Water Directive.[8] This value has been interpreted as the maximum allowable average concentration.
[b] Nickel is not one of the parameters covered by the Directive of 18th July 1978 on the quality of freshwaters needing protection or improvement in order to support fish life.[40]
[c] The maximum estimated intake from fish and shellfish is not thought to present a risk to human health.
[d] This standard is based on limited data for *Daphnia magna,* a species more normally associated with lacustrine environments.
[e] ADAS leaflet No. 776.[41] The value given is the suggested maximum average concentration during the normal irrigation period (i.e., the summer months). If exceeded, soil levels should be monitored.
[f] In order to make the standards for livestock watering consistent with others in Table 13, the "desirable" level has been interpreted as the annual average concentration. The "maximum tolerated level" might be as high as 50 mg/ℓ Ni and would apply to most stock.
[g] Other levels, to be derived locally, may be appropriate for individual industries and plants.
[h] This standard is based on the assumption that no more than one tenth of the water consumed by any individual is likely to be water accidentally swallowed during bathing or contact water sports. For most individuals, the consumption of water (and therefore nickel) by this mechanism would be negligible.
[i] Other values may be appropriate to local situations, e.g., for the protection of particularly sensitive species or shellfisheries.

Modified from Mance, G. and Yates, J., *Proposed Environmental Quality Standards for List II Substances in Water: Nickel,* TR211, Water Research Centre, Medmenham, U.K., 1984. With permission.

Table 14
EQS VALUES RECOMMENDED BY THE U.K. WATER RESEARCH CENTRE FOR ZINC

Use	EQS value (μg Zn/ℓ except where stated)			
	Average conc	95-percentile conc	Max allowable conc	Dissolved (D) or total (T)
Freshwater				
Direct abstraction to potable supply[a]				
Treatment Category				
A1	—	3,000*	4,500*	T
A2	—	5,000*	7,500*	T
A3				
Abstraction to a reservoir for potable supply[b]	—	$\dfrac{3{,}000 \times 100}{100 - R}$	—	T
Food for human consumption derived from freshwater[c]	—	—	—	—
Protection of freshwater fish[d]				
Salmonid fish				
Total hardness (mg CaCO$_3$/ℓ)				
10	8	30*	—	T
50	50	200*	—	T
100	75	300*	—	T
500	125	500*	—	T
Coarse fish				
Total hardness (mg CaCO$_3$/ℓ)				
10	75	300*	—	T
50	175	700*	—	T
100	250	1,000*	—	T
500	500	2,000*	—	T
Protection of other freshwater life and associated non aquatic organisms	100	—	—	D
Irrigation of crops[e]	1,000	—	—	T
Livestock watering[f]	25 mg/ℓ	—	—	T
Abstraction for food processing industry[g]	—	3,000	—	T
Bathing and contact water sports[h]	—	50,000	—	T
Saltwater				
Protection of saltwater fish and shellfish[i]	40	—	—	D
Protection of other saltwater life and[i] associated nonaquatic organisms	40	—	—	D
Bathing and contact water sports[h]	—	50,000	—	T
Food for human consumption derived from saltwater[c]	—	—	—	—

Note: * indicates a mandatory value. EQS, environmental quality standard.

[a] Required by the directive concerning the quality of surface water intended for the abstraction of drinking water in the member states.[7] Derogations allowed for floods, other natural disasters, and natural enrichment.

[b] R is the average extent of removal of zinc (percent) during passage through the reservoir. Water abstracted from the reservoir must comply with the Surface Water Directive.[7]

[c] The maximum estimated intake from fish and shellfish is not thought to present a risk to human health.

[d] Zinc is one of the determinands covered by the Directive of 18th July 1978 on the quality of freshwaters needing protection or improvement in order to support fish life.[40] Derogation is allowed for natural enrichment. EQS values for intermediate water hardnesses should be calculated by simple linear interpolation. Mandatory values apply to European Economic Communities designated fisheries only.

[e] ADAS leaflet No. 776.[41] The value given is the suggested maximum average concentration during the normal irrigation period (i.e., the summer months). If exceeded, soil levels should be monitored.

Table 14 (continued)
EQS VALUES RECOMMENDED BY THE U.K. WATER RESEARCH CENTRE FOR ZINC

f   In order to make the standards for livestock watering consistent with others in Table 14, the "desirable" level has been interpreted as the annual average concentration. A "maximum tolerated level" of 300 mg/ℓ has been suggested as suitable for most stock.
g   Other levels, to be derived locally, may be appropriate for individual industries and plants.
h   This standard is based on the assumption that no more than one tenth of the water consumed by any individual is likely to be water accidentally swallowed during bathing or contact water sports. For most individuals, the consumption of water (and therefore zinc) by this mechanism would be negligible.
i   Other values may be appropriate to local situations, e.g., for the protection of particularly sensitive species or shellfisheries,

Modified from Mance, G. and Yates, J., *Proposed Environmental Water Quality Standards for List II Substances in Water: Zinc*, TR209, Water Research Centre, Medmenham, U.K., 1984. With permission.

The philosophy behind the use of EQSs to control discharges has been extensively discussed in the literature.[43-48] In essence, monitoring is carried out at the point of protection and the required effluent standards specified on the basis of the prevailing local conditions. The degree of protection necessary is specified by the uses required of the particular body of water. This approach circumvents the requirement to identify all user industries together with the need to specify appropriate treatment technology for each user industry. The EQS approach does have certain disadvantages, however, which include a heavy reliance on our knowledge of the behavior of metals in the environment and in water treatment, in addition to bioaccumulation and toxic action. The data base on specific metals is usually incomplete, and indeed it is acknowledged that there will inevitably be a varying degree of uncertainty in the recommended EQS values.[49] Initial EQS values may subsequently be modified in the light of experience gained.

C. U.S. Government Criteria and Standards
1. Raw and Potable Waters

The EPA, formed in 1970, is generally responsible for the formulation and implementation of water-quality standards and the formulation of criteria, including those relating to potable water. Legislation concerning the quality of potable water is embodied in the Safe Drinking Water Act (SDWA) of 1974 as amended in 1977 and 1980, which is designed to maintain minimum uniform quality standards. The SDWA deals with drinking-water contamination problems through the creation of the Public Water System Supervision Program and the Underground Injection Control (UIC) Program.

The Public Water System Supervision Program is designed to protect public health by ensuring the safety of drinking water. This protection is based on three key elements: standards, implementation techniques, and enforcement responsibility. The SDWA requires EPA to promulgate primary drinking-water regulations which specify contaminants which may have an adverse effect on health, and for each contaminant specify either a maximum contaminant level (MCL) or treatment techniques. A treatment technique requirement would only be set "if it is not economically feasible or technologically feasible" to ascertain the concentration of a contaminant in drinking water. States which adopt and implement standards no less stringent than the national standards can obtain enforcement responsibility (primacy), while the EPA must implement a program in those states which do not have primacy.

As a result of the SDWA, the National Interim Primary Drinking Water Regulations (NIPDWR) were promulgated in 1975 and put in force in June 1977.[11] These regulations were referred to as "interim" since they were adopted to provide a means of assessing preliminary standards prior to the introduction of Revised National Primary Drinking Water

Table 15
MAXIMUM CONTAMINANT LEVELS FOR METALS INCLUDED IN THE NATIONAL INTERIM PRIMARY DRINKING WATER REGULATIONS[11]

Metal	Max contaminant level (mg/ℓ)
Ag	0.05
As	0.05
Ba	1.00
Cd	0.01
Cr	0.05
Hg	0.002
Pb	0.05
Se	0.01

Table 16
SECONDARY MAXIMUM CONTAMINANT LEVELS FOR METALS INCLUDED IN THE NATIONAL SECONDARY DRINKING WATER REGULATIONS[50]

Metal	Secondary max contaminant level (mg/ℓ)
Cu	1.0
Fe	0.3
Mn	0.05
Zn	5.0

Regulations (RNPDWR).[13] The term "primary" is used to denote those substances which are known or suspected to exert adverse effects on health. Under the NIPDWR, enforceable standards were set in terms of MCLs, together with monitoring and reporting requirements for eight metals (Table 15).

In the RNPDWR "recommended maximum contaminant levels" (RMCLs) must also be specified. RMCLs are nonenforceable health goals for public water systems which are to be set at a concentration at which "no known or anticipated adverse effects on the health of persons occur and which allows an adequate margin of safety". In the case of carcinogens where no safe exposure dose can be demonstrated (since any exposure represents a finite level of risk), the RMCL "should be set at zero level".[13] In the RNPDWR, MCLs must be set as close to RMCLs as is feasible, taking into account:

1. Potential health risks
2. Performance of available treatment technologies
3. Feasibility and costs of treatment
4. Analytical methods: levels of precision and accuracy attainable by qualified laboratories

The SDWA also requires that the RNPDWR be reviewed every 3 years and amended whenever changes in technology, treatment techniques, or other factors permit greater health protection. Under certain circumstances, variances may be issued to provide legal protection to systems that are unable to comply with the regulations, despite the application of treatment technologies, because of poor source quality.

In addition to the primary regulations, the SDWA requires the EPA to set National Secondary Drinking Water Regulations (NSDWR) for those substances considered to have no adverse effect on health but which may impart organoleptic properties to drinking waters. NSDWR were established by July 1979[50] and introduced secondary maximum contaminant levels (SMCLs) (Table 16) and monitoring requirements. These are intended to act as guidelines for state regulatory bodies rather than being federally enforceable.

A major part of EPA efforts to protect underground sources of drinking water is in the UIC program. EPA promulgated the UIC regulations in 1980,[14] under which the states are required to demonstrate that their programs protect underground sources of drinking water.

The Federal Water Pollution Control Act Amendments of 1972 require the EPA to publish criteria for water quality "accurately reflecting the latest scientific knowledge on the kind and extent of all identifiable effects on health and welfare which may be expected from the presence of pollutants in any body of water, including ground water". This resulted in the publication by EPA of *Quality Criteria for Water* (the "Red Book") due in 1973 but published in 1976. The "Red Book", which was intended to "recommend criteria levels for a water quality that will provide for the protection and propagation of fish and other aquatic life and for recreation in and on the water", specified values based on chronic, lifetime bioassays or extrapolation equivalent of sensitive organisms exposed to contaminants in 100% available forms, portraying the "worst case" criteria.[51] Subsequently, the criteria were amended and republished in 1980.[52]

Following the enactment of the Clean Water Act in 1977, and in satisfaction of the "Settlement Agreement" in Natural Resources Defense Council et al. vs. Train of 1976, the EPA was required to periodically review and publish criteria documents for the 65 pollutants and classes of pollutants which Congress designated as toxic.[53] In November 1980, criteria documents were issued for 64 of the 65 pollutants and classes including all 13 heavy metals specified.[14] Although this list of pollutants is extensive, certain criteria specified in the "Red Book" have not been replaced, and these remain valid. It is stressed that these criteria are not mandatory requirements but merely present scientific data and guidance on the environmental effects of pollutants which can be useful to derive regulatory requirements based on considerations of water-quality impacts. Criteria relating to heavy metals contained in these documents and those remaining valid from the "Red Book" are summarized in Table 17.

Criteria published in 1980 were derived using revised methodological "guidelines" for calculating the impact of pollutants on human health and aquatic organisms. These "guidelines" consist of "systematic methods for assessing valid and appropriate data concerning acute and chronic adverse effects of pollutants on aquatic organisms, nonhuman mammals, and humans. By using these data in certain ways, criteria were formulated to protect aquatic life and human health from exposure to the pollutants".[14] Individual criteria documents contain information relating to two specific aspects: discussions of available scientific data on the effects of pollutants on public health and welfare, aquatic life, and recreation, and quantitative concentrations or qualitative assessments of the pollutants in water which will generally ensure water quality adequate to support a specific water use.

Criteria designed to protect aquatic life are not intended to provide complete protection of all species and all uses of aquatic life all of the time, but rather to protect most species in a balanced, healthy aquatic community. The criteria values were derived from a consideration of four specific areas each based on at least a set minimum data base:

1. **Final acute value** — Based on data for fish and invertebrate species designed to protect most, but not necessarily all, of the tested and untested species.
2. **Final chronic value** — Determined in a similar manner where data is available or based on acute-chronic toxicity ratio.
3. **Final plant value** — Based on lowest aquatic plant toxicity value determined by decreased growth rate.
4. **Final residue value** — Calculated by dividing a maximum permissible tissue concentration by an appropriate bioconcentration factor. Designed to protect commercially and recreationally important aquatic organisms in addition to wildlife which consume aquatic organisms.

Criteria maximum values quoted in Table 17 are equivalent to final acute values while the lower of the remaining values sets the 24-hr average concentration. If sufficient data are available to demonstrate that one or more of the final values should be related to a water-quality characteristic, e.g., hardness for silver toxicity in freshwater (Table 17), the final value is expressed as a function of that characteristic.

It is acknowledged by the EPA that the criteria are based on the effects of pollutants in

Table 17
WATER QUALITY CRITERIA FOR METALS ACCORDING TO THE EPA[14,52]

Metal	Form	Basis	Limit for freshwater aquatic life ($\mu g/\ell$)	Limit for saltwater aquatic life ($\mu g/\ell$)	Limit for human health ($\mu g/\ell$)
Ag	T[a]	max	H[b]	2.3	50[c]
As	As^{3+}	max	440	—	0.0022[d]
Ba	T	max	—	—	1000[c]
Be	T	max	—	—	0.0037[d]
Cd	T	24 hr av	H[e]	4.5	10[c]
		max	H[f]	59	
Cr	Cr^{6+}	24 hr av	0.29	18	
		max	21	1260	50[c]
	Cr^{3+}	max	H[g]	—	170
Cu	T	24 hr av	5.6	4.0	1000[i]
		max	H[h]	23	
Fe	T	max	1000	—	300[i]
Hg	T	24 hr av	0.00057	0.025	
		max	0.0017	3.7	0.144
Mn	T	max	—	100	50[i]
Ni	T	24 hr av	H[j]	7.1	
		max	H[k]	140	13.4
Pb	T	24 hr av	H[l]		
		max	H[m]	—	50[c]
Sb	T	max	—	—	146
Se	T	24 hr av	35	54	
		max	260	410	10[c]
Tl	T	max	—	—	13
Zn	T	24 hr av	47	58	5000[i]
		max	H[n]	170	

[a] Total recoverable; criteria related to water hardness (mg/ℓ $CaCO_3$).
[b] e [1.72 (ln hardness) − 6.52].
[c] Numerically equal to existing drinking-water standard.
[d] Value based on risk estimate of one additional cancer case per million of population.
[e] e [1.05 (ln hardness) − 8.53].
[f] e [1.05 (ln hardness) − 3.73].
[g] e [1.08 (ln hardness) + 3.48].
[h] e [0.94 (ln hardness) − 1.23].
[i] Organoleptic related.
[j] e [0.76 (ln hardness) + 1.06].
[k] e [0.76 (ln hardness) + 4.02].
[l] e [2.35 (ln hardness) − 9.48].
[m] e [1.22 (ln hardness) − 0.47].
[n] e [0.83 (ln hardness) + 1.95].

isolation and that synergistic, additive, and antagonistic properties are not considered. However, the criteria are designed to be flexible in nature and may be modified by individual states to reflect local environmental conditions and human exposure patterns prior to incorporation into water-quality standards. In particular, criteria developed for the protection of aquatic life could be determined using acute and chronic toxicity data for local species or may be specifically tailored to a local water body by performing toxicity tests using the local ambient water. Furthermore, site-specific factors may be included in the criteria development or more stringent criteria may be advanced to protect an individual species of local importance.

Human health criteria have been developed by the EPA using a combination of epidemiological and animal dose/response data. No effect (noncarcinogenic) or specified risk (carcinogenic) concentrations were estimated by extrapolation to assess the relative extent of human exposure either directly through ingestion of water or indirectly via consumption of aquatic organisms. Health criteria for carcinogens are presented as incremental risks to man associated with specific concentrations in ambient water (Table 17). Where no carcinogenic effect is apparent, criteria are based on acceptable daily intake (ADI) levels and are generally derived using no-observable-adverse-effect levels data from animal studies although human data are used wherever available. Unfortunately the definition of a no-effect level may range from no gross effects such as mortality, to more subtle biochemical, physiological, or pathological changes, and the ADI is calculated using safety factors to account for uncertainties inherent in extrapolation from such animal data to man. The development of human health criteria for water is not intended to replace drinking-water legislation in the U.S., and the criteria have no regulatory significance. However, the criteria may be considered as target levels for contaminants in drinking water at which "no known or anticipated adverse effects occur and which allows an adequate margin of safety".[14] It should be noted that the criteria do not consider treatment technology or the economics of drinking-water treatment, nor do they take into account exposure from other sources.

Draft revisions of certain water-quality criteria have been published by the EPA which propose major technical changes in the "guidelines" for deriving numerical water-quality criteria for the protection of fresh- and salt-water aquatic life.[54] In this draft document, the acute toxicity data required for freshwater animals has been amended to include more tests with invertebrate species, while final acute value is defined in terms of "family" rather than "species" mean acute values. A family mean acute value is the geometric mean of all the species acute values available for species in the family. The proposed criteria consist of two numbers: the criterion average concentration based on 30 consecutive days monitoring and the criterion maximum concentration. Excursions above the average are to be limited to one 96-hr period in any 30 days. It is further proposed that the criterion maximum concentration be obtained by dividing the final acute value by two. Changes in toxicity testing are also envisaged. These modifications would result in the ambient water-quality criteria for arsenic, cadmium, chromium, copper, mercury, and lead being revised as indicated in Table 18.

2. Wastewaters

Under the Federal Water Pollution Control Act Amendments of 1972, existing dischargers of wastewater were required to achieve "effluent limitations requiring the application of the best practicable control technology currently available (BPT) by July 1, 1977. By July 1, 1983 these same dischargers were required to achieve BAT defined as "effluent limitations requiring the application of the best available technology economically achievable". New industrial dischargers were required to comply with new source performance standards (NSPS) based on best available demonstrated technology while new and existing dischargers to publicly owned treatment works (POTW) (i.e., indirect dischargers) were subject to pretreatment standards. Direct dischargers were to be incorporated into the National Pollutant Discharge Elimination System (NPDES). The act required EPA to promulgate regulations for effluent quality through the application of BPT and BAT together, with the promulgation of regulations for NSPS, pretreatment standards for existing sources (PSES), and pretreatment standards for new sources (PSNS). Moreover, the act required the EPA to develop a list of toxic pollutants and promulgate effluent standards applicable to all dischargers of these pollutants.[20]

The EPA failed to promulgate many of the regulations by the dates required, which resulted in a consent decree being issued by the U.S. Courts following the result of Natural

Table 18
PROPOSED REVISED WATER QUALITY CRITERIA FOR METALS ACCORDING TO THE EPA[54]

Metal	Form	Basis	Limit for freshwater aquatic life ($\mu g/\ell$)	Limit for marine aquatic life ($\mu g/\ell$)
As	As^{3+a}	30 day av	72	63
		30 day max[b]	140	120
Cd	(Footnote[c])	30 day av	—	12
		30 day max[b]	—	38
		max	H[d]	—
Cr	Cr^{6+a}	30 day av	7.2	54
		30 day max[b]	11	1200
	Cr^{3+c}	30 day av	H[e]	—
		30 day max[b]	H[f]	—
Cu	(Footnote[c])	30 day av	H[g]	2.0
		30 day max[b]	H[h]	3.2
Hg	Hg^{2+c}	30 day av	0.2	0.1
		30 day max[b]	1.1	1.9
Pb	(Footnote[c])	30 day av	H[i]	8.6
		30 day max[b]	H[j]	220

[a] Dissolved; passes through a 0.45-μm membrane filter.
[b] Concentration may be between average and maximum values for up to 96 hr/30 days.
[c] Active; passes through a 0.45-μm membrane filter following acidification to pH 4 with nitric acid. Criteria related to water hardness (mg/ℓ CaCO$_3$).
[d] $e[1.16 (\ln \text{hardness}) - 3.841]$.
[e] $e[0.819 (\ln \text{hardness}) + 0.537]$.
[f] $e[0.819 (\ln \text{hardness}) + 3.568]$.
[g] $e[0.905 (\ln \text{hardness}) - 1.785]$.
[h] $e[0.905 (\ln \text{hardness}) - 1.413]$.
[i] $e[1.34 (\ln \text{hardness}) - 5.245]$.
[j] $e[1.34 (\ln \text{hardness}) - 2.014]$.

Resources Defense Council, Inc. vs. Train in 1976. Several of the basic elements of the consent decree were incorporated in the Clean Water Act of 1977 which identified 21 point source categories and required EPA to promulgate BAT effluent-limitation guidelines, NSPS, PSES, and PSNS for 65 "priority pollutants" and classes of pollutants including the following metals and their compounds: antimony, arsenic, beryllium, cadmium, chromium, copper, lead, mercury, nickel, selenium, silver, thallium, and zinc.[53] Regulations concerning direct discharges (not via POTWs) are applied to specific industrial plants under the NPDES permit program. These permits are issued by the EPA, or the state director of an approved state program.[50] To control the introduction of wastes from industry into POTWs, the EPA has promulgated General Pretreatment Regulations for Existing and New Sources. These regulations have, however, subsequently been amended and in part deferred following considerable discussion and several lawsuits.[55]

Regulations governing the discharge of toxic metals and metal compounds into waters

Table 19
METAL LIMITATIONS ACCORDING TO THE EPA FOR DISCHARGES BY THE CHLOR-ALKALI ELECTROLYSIS INDUSTRY[17]

Process[a]	Metal	BPT (g/tonne)[b]		BAT (g/tonne)[b]		PSES (mg/ℓ)[c]		NSPS (g/tonne)[b]		PSNS (mg/ℓ)[c]	
		Max	Mean	Max	Mean	Max	Mean	Max	Mean	Max	Mean
Mercury cell	Hg	0.28[d]	0.14	0.23[e]	0.10	(Footnote[f])		0.23	0.10	0.11	0.048
Diaphragm cell	Cu	18[g]	7	12[h]	4.9	2.1[i]	0.8				
	Ni	14	5.6	9.7	3.7	1.6	0.64				
	Pb	26	10	5.9	2.4	2.9	1.1	4.7[j]	1.9	0.53[k]	0.21

Note: BPT, effluent limitations, requiring the application of the best practicable control technology currently available. BAT, effluent limitations requiring the application of the best available technology economically achievable. PSES, pretreatment standards for existing sources. NSPS, new source performance standards. PSNS, pretreatment standards for new sources. Max, maximum for any one day. Mean, average of daily values for 30 consecutive days.

[a] Processes differ in cell design and in quantity and quality of wastewater generated.
[b] Grams discharged per tonne of chlorine production.
[c] Milligrams discharged per liter of effluent.
[d] Specifies sulfide precipitation and filtration of mercury-laden streams; neutralization and settling of brine muds.
[e] As in previous footnote plus dechlorination.
[f] Excluded, since discharges below treatable levels.
[g] Specifies neutralization, alkaline precipitation, and settling.
[h] As in previous footnote plus dual media filtration and dechlorination.
[i] Mass discharge limitations are equal to BPT.
[j] Achieved by change in industrial process.
[k] Mass discharge limitations are equal to NSPS.

and POTWs by existing and potential new sources which produce certain inorganic chemicals have been revised and promulgated by the EPA as "Inorganic Chemicals Manufacturing Point Source Category Effluent Limitations Guidelines, Pretreatment Standards, and New Source Performance Standards", which came into effect on August 12, 1982.[17] This document details the effluent-treatment options available for these plants and provides an account of the technological basis for the final regulations specified. Under this regulation, and subject to the articles of the NPDES regulations, BPT was enforceable while BAT limitations were due to be in force by July 1, 1984. NSPS and PSNS are applicable to all new sources as appropriate, while PSES are applicable to all dischargers to POTWs from August 12, 1985. An evaluation of the effect of pollution control costs on these industries concluded that the economic impact would not be significant and would be justified in view of the benefits associated with compliance.

Effluent limitations for industrial processes amended by these regulations include those for the chlor-alkali electrolysis industry together with the production of aluminum fluoride, chrome pigments, copper sulfate, hydrogen cyanide, hydrofluoric acid, nickel sulfate, sodium bisulfite, sodium dichromate, and titanium dioxide. Specific exclusion of certain effluent limitations (e.g., BPT, PSNS, or PSES) are detailed together with the metals to be controlled in the effluents from each process. Specific toxic pollutants may be excluded from regulations applying to each industrial category if the pollutant is undetectable in the effluent or is present in trace amounts not likely to cause toxic effects. Alternatively an exclusion may be granted if the pollutant is present in amounts too small to be effectively reduced by existing treatment technology or if the pollutant is effectively controlled by the technology

Table 20
PROPOSED METAL LIMITATIONS ACCORDING TO THE EPA FOR DISCHARGES BY THE CADMIUM PIGMENTS AND CADMIUM SALTS PRODUCTION INDUSTRIES[20]

Industrial sector[a]	Metal	BPT, BAT, NSPS (g/tonne)[b]		PSES, PSNS (mg/ℓ)[c]	
		Max	Mean	Max	Mean
Pigments	Cd	43[d]	14	0.47[d]	0.15
	Se	110	37	1.1	0.40
	Zn	11	6.2	0.12	0.067
Salts	Cd	0.0027[d]	8.7×10^{-5}	0.47[d]	0.15
	Se	0.070	0.023	1.1	0.40
	Zn	0.007	3.9×10^{-3}	0.12	0.067

Note: BPT, effluent limitations requiring the application of the best practicable control technology currently available. BAT, effluent limitations requiring the application of the best available technology economically achieveable. NSPS, new source performance standards. PSES, pretreatment standards for existing sources. PSNS, pretreatment standards for new sources. Max, maximum for any one day. Mean, average of daily values for 30 consecutive days.

[a] Processes differ in quantity of wastewater generated.
[b] Grams discharged per tonne of production.
[c] Milligrams discharged per liter of effluent. Mass discharge limitations are equal to BPT.
[d] Specifies lime precipitation with clarification followed by filtration.

upon which other limitations are based. In particular, antimony, arsenic, beryllium, cadmium, silver, and thallium and their compounds are excluded from all regulations applying to these industries on the grounds that they are present in the effluents at concentrations below the level of treatability. Certain industrial categories are deferred to later regulation, or excluded under various articles of the consent decree.[17] It is beyond the scope of this chapter to consider each industrial category with regard to the discharge of toxic metals, but in order to facilitate comparison, the regulations relating to the chlor-alkali electrolysis process are specified in Table 19. A proposed rule has been published by the EPA which amends further articles of the discharge regulations for various subcategories of the inorganic chemicals industry.[20] An outline of the proposed regulations for cadmium, selenium, and zinc in discharges from the cadmium pigments and salts industries is included in Table 20.

3. Other Classes of Water

In addition to the reuse of water for potable supply, reclaimed water may be exploited for a variety of purposes including agricultural and industrial use. In the U.S., water-quality requirements have been stipulated for the use of reclaimed water in agricultural irrigation, livestock watering, industrial processes, and groundwater recharge, thereby determining the type and extent of wastewater treatment necessary for a specific water use.[56] Irrigation represents the single largest and most widely adopted example of controlled direct reuse of water in the U.S. Water-quality criteria for agricultural irrigation and groundwater recharge are therefore presented in Table 21.[56] Under the U.S. guidelines direct injection of wastewater

Table 21
WATER QUALITY CRITERIA FOR AGRICULTURAL IRRIGATION AND GROUNDWATER RECHARGE IN THE U.S.[56]

Metal	Criteria for continuous agricultural irrigation (mg/ℓ)	Criteria for groundwater recharge (mg/ℓ)
Ag	—	0.10
Al	5	—
As	0.10	0.05
Ba	—	2
Be	0.10	—
B	1	0.02
Cd	0.01	0.01
Co	0.05	—
Cr	0.10	0.15
Cu	0.20	2
Fe	5	0.10
Hg	—	0.01
Li	2.5[a]	—
Mn	0.20	0.10
Mo	0.01	—
Ni	0.20	—
Pb	5	0.05
Se	0.02	0.01
V	0.10	—
Zn	2	10

[a] Value is 0.075 mg/ℓ for citrus crops.

into groundwater requires the application of drinking-water standards, while groundwater recharge by percolation through the soil profile must be carried out applying the criteria presented in Table 21. Such differences are based on the premise that the movement of heavy metals in the soil profile, following the continuous application of secondary effluent, is limited.[57]

D. Water-Quality Standards Issued by the Russian Ministry of Health
1. Potable Waters

The most recent criteria for drinking water in the U.S.S.R. were issued in 1973[12] and have been reviewed by Cherkinskiy.[58] Limits for metals in drinking water based on toxicological and organoleptic properties are listed in Table 22. When a number of toxic substances (including those toxic metals specified in Table 22) are detected in water, the total concentration, expressed in fractions of the maximum tolerable concentration of each separate substance should not exceed unity. Similarly, for organoleptic substances (including those metals specified in Table 22), their total concentration, expressed in fractions of the maximum tolerable concentration of each separate substance, should not exceed unity.

2. Raw Waters

Comprehensive lists of water-quality standards for surface and groundwaters intended for the production of water for hygienic and domestic purposes were published in 1970[59] and

Table 22
DRINKING WATER STANDARDS FOR METALS ESTABLISHED BY THE U.S.S.R. MINISTRY OF HEALTH[12]

Metal	Standard (mg/ℓ)
Ag[a]	0.05
As	0.05
Be	0.0002
Pb	0.1
Mo	0.5
Se	0.001
Sr	2.0
Cu[b]	1.0
Fe[b]	0.3
Mn[b]	0.1
Zn[b]	5.0

[a] When used for water preservation.
[b] Organoleptically based.

were reviewed by Stöfen.[60] Three categories of regulations were issued: toxicological limits based upon reflexological toxicity testing, limits for substances which affect the self purification of water courses, and organoleptic limits. Metals, not including those listed as metalloorganic compounds, considered in the lists and their respective limit concentrations are included in Table 23. When the water in water courses used for domestic purposes is polluted by a combination of different substances which belong to the same category of hazard, i.e., toxicological quality, impairment of self-purification or organoleptic properties, the maximum permissible concentrations in Table 23 for a single substance must be adopted under one of the following provisions.[59]

1. In carrying out preventative sanitary inspection, the maximum permissible concentration for each substance forming the combination must be reduced as many times as the number of hazardous substances within the same category which are supposed to be discharged with wastewater or are contained in the water course.
2. In carrying out routine sanitary inspection, the sum of the concentrations of all the substances expressed as a percentage of the corresponding maximum permissible concentration for each substance separately must not exceed 100%.

Therefore, although synergistic and other interaction effects between metals are not specified, the provisions of the standards do take some account of combinations of toxic and organoleptic pollutants. The standards have been criticized, however,[60] particularly with regard to the failure to take into account special problems of infants and children associated

Table 23
MAXIMUM PERMISSIBLE CONCENTRATIONS OF METALS IN THE WATER OF WATERCOURSES USED FOR HYGIENIC AND DOMESTIC PURPOSES IN THE U.S.S.R.[59]

Metal	Max permissible conc (mg/ℓ)
As[a]	0.05
Be[a]	0.0002
Co[a]	1.0
Hg[a]	0.005
Mo[a]	0.5
Pb[a]	0.1
Sb[a]	0.05
Se (SeO$_3$)[a]	0.001
Sr[a]	2.0
Te[a]	0.01
V[a]	0.1
W[a]	0.1
Cd[b]	0.01
Cu[b]	0.1
Ni[b]	0.1
Ti[b]	0.1
Zn[b]	1.0
Fe[c]	0.5
Cr (Cr^{6+})[c]	0.1
Cr (Cr^{3+})[c]	0.5

[a] Toxicologically based.
[b] Substances which affect the self purification of water courses.
[c] Organoleptically based.

with their higher metabolic rate and therefore higher intake of drinking water relative to body weight. It is also apparent that no standards relating to drinking water were issued for cadmium and mercury, although standards for these two metals were included in the limits for water courses used for potable abstraction.

III. SLUDGE DISPOSAL TO LAND

Due to the often high concentrations of heavy metals in sludge, many countries have developed legislation or recommendations which specify the maximum concentrations of metals in sludge which may be applied to agricultural land. In general, total concentrations are specified, and metal speciation is not taken into account, with few exceptions. In the U.S., the quantities of metals which may be applied to land depend on the cation-exchange capacity of the soil, the assumption being that a soil with a cation exchange capacity of $>$ 15 mg/100 g can tolerate a metal load four times greater than one with a cation-exchange capacity of $<$ 5 mg/100 g.[61] The indirect effect of pH on metal availability is also taken into account in the recent CEC proposal[25] which states that soil pH should be maintained above 6.0, and in the U.K. guidelines[24] which recommend a pH of 6.0 for grassland, 6.5 for arable crops, and $>$ 7.0 for significantly contaminated soil.

Certain authorities have taken plant-uptake factors into account when specifying limits.

The Sewage Sludge Ordinance in Germany states that sludge must not be applied to grassland, horticultural, or fruit crops during the vegetative period up to the date of harvesting,[62] while in Holland, different quantities of sludge are specified depending on whether it is applied to arable land or pasture.[62] In the U.K., the permissible additions of the phytotoxic metals zinc, copper, and nickel may be doubled if the sludge is to be applied to grassland rather than arable land, although reservations about this have recently been made.[24]

The potential interactive effects of heavy metals has been recognized in the U.K. by the incorporation into the guidelines of the zinc-equivalent concept.[22] This supposes that copper is twice, and nickel eight, times as toxic as zinc on a weight basis, and that the toxicities are additive. However, recent evidence suggests that the toxic effects of these elements only become additive when one or more approaches its individual toxic concentration.[63-65] The validity of the zinc-equivalent concept will depend on the available concentrations of heavy metals, rather than their total concentrations. The toxicities of copper and nickel to red celery and lettuce were found to be less than twice and less than three times, respectively, as toxic as zinc when based on total soil concentrations, but based on soluble concentrations were 0.4 and 5.9 times as toxic as zinc, respectively.[66]

The specification of limitations on heavy-metal additions is quite variable. In Germany, maximum tolerable concentrations of heavy metals in the soil are defined,[67] while in Holland,[68] Finland, and Denmark,[69] limitations are placed on heavy-metal concentrations in the sludge. In the latter country, total sludge addition is based on the crop nitrogen demand, whereas maximum upper limits of sludge addition are specified in addition to limits on heavy-metal concentrations in Sweden, Norway, and Holland. In some cases, both the maximum tolerable concentrations of metals in the sludge and the limits on the concentrations of metals to be applied to land are specified.[69]

The limitations on heavy-metal additions in sludge to soil in several countries are shown in Table 24. Annual averages are given in each case, although in the U.K. limitations are based on a period of "30 years or more". However if sludges are not applied every year, the maximum quantity which may be applied in any one year is six times the annual average.[64]

An example of the calculation of the quantity of sludge (dry solids) which may be applied to land over 30 years is given below. The application limit in respect of each metal is given by:

$$\text{Limit (t/ha)} = \frac{\text{recommended maximum addition of metal (kg/ha)} \times 1000}{\text{metal concentration in sludge (mg/kg)}}$$

Assuming, for example, that a sludge contains the following concentrations of metals (mg/kg): cadmium, 25; chromium, 500; copper, 600; nickel, 300; lead, 1000; and zinc, 2000 and negligible concentrations of other metals, then the limits of application for each metal over 30 years are

$$\begin{aligned}
\text{cadmium} &= (5 \times 1000)/25 = 200 \text{ t/ha} \\
\text{chromium} &= (1000 \times 1000)/500 = 2000 \text{ t/ha} \\
\text{lead} &= (1000 \times 1000)/1000 = 1000 \text{ t/ha} \\
\text{zinc equivalent} &= (560 \times 1000)/2000 + (2 \times 600) + (8 \times 300) \\
&= 100 \text{ t/ha}
\end{aligned}$$

If the sludge were to be applied annually, then the limit of addition would be 3.3 t/ha/year (ie., 1/30 of the smallest figure obtained from the above calculations). If applied periodically, the maximum quantity which could be applied in any one year would be one fifth of the total, i.e., 20 t/ha. This calculation applies to the application of sludge to uncontaminated soils.[24] Where soils are contaminated, the quantities of sludge to be applied should be reduced so that maximum soil concentrations (Table 24) are not exceeded after 30 years.

Table 24
LIMITATIONS IMPOSED ON HEAVY METALS IN SLUDGE APPLICATION TO LAND[70]

Metal	Limits of metal addition to land (annual av) kg/ha				Total kg/ha	Limits of metal conc in sludge (mg/kg)			Limits of metal conc in soil	
	U.S.[a]	U.K.	EEC	Sweden	Norway	Holland	Finland	Denmark	FRG[f] (mg/kg)	U.K.[g] (mg/ℓ)
As									20	10
B									25	3.25
Be									10	
Cd	0.17	0.33	0.35[b]	0.015		10	30	30[e]	5	3.5
Cr		3.5		1.0	30	500	1000	500	100	600
Co		0.17		0.05			100		50	
Cu	4.17	33.3	0.15	3.0		500	3000	700	100	280
Hg		9.3	10[b]	0.008	14	10	25		5	1
Mo		0.07	12						5	4
Ni	1.7	0.13		0.5		50	500		50	70
Pb	16.7	2.3	0.40[b]	0.3	600	500	1200	1200	100	550
Sb		33.3	3						5	
Se		0.17	15						10	3
Sn									50	
V									50	
Zn	8.3	18.7	30	10.0		2000	5000	6000	300	560
Period of addition (years)	30	30	20	Single addition[c]		5 annual dosings				
Limit of addition of sludge (t/ha/year)			1		2[d]	2 (arable) 1 (pasture)		Calculated from N demand		

Note: EEC, European Economic Communities. FRG, Federal Republic of Germany.

a These limits apply if soil cation exchange capacity (CEC) is < 5 mg/100g.
b Recommended values. Other values given are mandatory.
c Repeated application is discouraged.
d Maximum sludge application is 50t/ha.
e Cadmium limit is 0.015 kg/ha/year.
f Tolerable concentrations. (mg/kg = mg/ℓ)
g Calcareous soils. Limits for zinc equivalents in noncalcareous soils are halved.

Table 25
HEAVY METALS LIMITING THE APPLICATION OF 40 SLUDGES TO LAND IN THE U.K.[71]

Limiting metal	Sludges limited (%)
Cd	10
Cr	0
Mo	5
Pb	2.5
Zinc equivalent	82.5

Table 26
THEORETICAL MAXIMUM ANNUAL AVERAGE RATES OF ADDITION OF 40 SLUDGES TO LAND[71]

Rate of addition (t/ha/year)	Percentage of sludges
<2	22.5
2—5	35
6—10	27.5
>10	15

These calculations were applied to the data obtained by Sterritt and Lester[71] for 40 sludges in England, and the frequencies of limitations by the zinc equivalent, cadmium, molybdenum, and lead are shown in Table 25. From these data, it would appear that under present criteria, the concentrations of copper, nickel, and zinc would most frequently limit sludge applications to land. The theoretical limits of addition were also calculated individually for each sludge from the survey[71] and are summarized in Table 26. It is interesting to note that these acceptable quantities are in stark contrast to the additions of 62 t/ha applied annually in a long-term study of the agricultural use of sewage sludge[72] and the 450 t/ha applied in a study of the phytotoxic effects of metals.[73] One important consideration in the agricultural use of sludge is the nitrogen requirement of crops. An acceptable addition of sludge to fulfil this requirement would be about 3 to 4 t/ha. It is therefore apparent that a significant proportion of sludges would be unsuitable for application to agricultural land. This proportion is likely to increase as recommended limits are made more stringent. Meixner[74] found that in Bavaria, at least 30% of the sludges produced would fail to comply with the latest regulations for land disposal under the terms of the Sewage Sludge Ordinance. Ultimately, therefore, alternative sites for disposal of these sludges will have to be found.

IV. SLUDGE DISPOSAL TO SEA

Sludge disposal to sea in the U.K. accounted for 29% of sludge production in 1977[24], and approximately 45% of this total arose from Thames Water Authority disposals in the outer Thames Estuary.[75] On an international basis, sludge disposal to sea has been the subject of considerable debate and the U.S. legislated to phase out the ocean disposal of digested sewage sludge by 1982[28] in favor of alternative methods, although a reprieve was granted to the City of New York who won a ruling in the Federal District Court which may allow continued disposal for several years.[76]

Although the sea has a finite capacity to accept certain quantities of toxic wastes without suffering any ill effects to its flora and fauna, due regard must be given to the method of disposal, the rate of dilution and dispersion, and the possible mechanisms for aggregation of the waste or its reconcentration by biological systems.[27] As a result of this concern, two international conventions were agreed to in 1972, and have since been ratified, to control the dumping of wastes from vessels. The Oslo Convention,[77] (Convention for the Prevention of Marine Pollution by Dumping from Ships and Aircraft), which has 13 states as signatories, applies to the North Sea and northeast Atlantic, while the London Convention[78] (Final Act of the Intergovernmental Conference on the Convention of Dumping of Wastes at Sea) has 60 states as signatories and applies globally.

In both conventions, substances considered for disposal are listed in groups. In List 1, substances such as mercury, cadmium, and their compounds are prohibited. In List 2, substances such as copper, lead, zinc, and arsenic require special permits. In List 3, substances, including sewage sludge, are controlled by general permits which cover a number of factors including the location.

Sewage sludges contain List 1 substances which are prohibited and List 2 substances, but they still qualify as List 3 substances if the List 1 and List 2 substances are present at "trace" concentrations.[79] "Trace" concentrations of List 1 substances have been interpreted by the Ministry of Agriculture, Fisheries, and Food (MAFF) to represent limits of 40 mg/kg (dry solids) of cadmium and 20 mg/kg (dry solids) of mercury.[80] In a survey of 193 sewage-treatment works in 1977, 25% of the sludges from those works contained in excess of 40 mg/kg (dry solids) of cadmium[24], and it would appear from data presented in 1978 that these limits for cadmium and mercury may have been exceeded in the Thames Estuary and in the Liverpool Bay, respectively.[81] The dumping of List 2 substances is strictly controlled where the concentrations of these substances exceed 0.1% by weight of the sludge.[27] Under the terms of the conventions, any licensing authority must notify the commissions (and hence all parties of the conventions) immediately of any dumping operations involving List 2 substances exceeding these concentrations. Licenses for List 3 substances are notified every 3 months, having due regard to the composition and properties of the waste, characteristics of the proposed dumping area, and method of disposal.

U.K. legislation to control the dumping of waste, including sewage sludge, at sea is incorporated in the 1974 Dumping at Sea Act[26] under which it is an offence to dump (or load for the purpose of dumping) any material in the sea, from a vehicle, ship, aircraft, hovercraft, or other marine structure without a licence from the relevant licensing authority, who are charged with the protection of the receiving environment and the monitoring of the dumping ground and nature of the waste. The provisions of the act thereby encompass the rationale behind List 3 of the Conventions, although a detailed schedule of List 1 and List 2 substances is not included in the U.K. legislation.[27]

Discharges from "land-based sources" by the European states bordering the North Atlantic are regulated by the Paris Convention[82] (Convention for the Prevention of Pollution from Land-based Sources) to which the U.K. is a signatory and has now ratified. The form of the Paris Convention is similar to that of the Oslo and London Conventions in that certain substances (including cadmium, mercury, and their compounds) are listed in Part 1 as substances whose discharges should be eliminated, while Part 2 lists substances (including arsenic, chromium, copper, lead, nickel, and zinc) whose discharge should be strictly controlled. In addition to the Paris Convention, those countries which are members of the European community are bound by the Dangerous Substances Directive[32] which applies to both fresh and marine waters. The directive intends to limit the discharge of certain substances by specifying limit values on emissions of specific substances to be specified in supplementary directives.

The 1974 Control of Pollution Act[83] provides much wider powers and, when implemented, will bring under water-authority control all discharges to coastal waters. However, as yet the U.K. is not in a position to meet all of its obligations to either the Paris Convention or any subdirective of the Dangerous Substances Directive.[27]

V. COMPARISON OF WATER QUALITY FOR IRRIGATION AND SLUDGE DISPOSAL GUIDELINES

Contrasting the criteria for agricultural irrigation and groundwater recharge, it is apparent that phytotoxic elements such as copper, nickel, and zinc require greater control for agricultural irrigation. However, limits for continuous agricultural irrigation appear to be less stringent than those adopted for sewage sludge disposal to agricultural land in the U.S. Table 27 includes the volume of wastewater effluent which would be required to achieve the limit

Table 27
COMPARISON OF LIMITS FOR THE APPLICATION OF CERTAIN METALS TO AGRICULTURAL LAND IN THE U.S. VIA IRRIGATION AND SEWAGE SLUDGE DISPOSAL[56,61]

Metal	Limit (A) for metals in agricultural irrigation (mg/ℓ)	Limit (B) for annual metal addition in sewage sludge (mg/m^2)	Irrigation volume at limit (A) required to reach limit (B) per annum (ℓ/m^2)
As	0.10	(Footnote[a])	—
Cd	0.01	17	1700
Cu	0.20	417	2085
Ni	0.20	170	850
Pb	5	1670	334
Zn	2	830	415

[a] No national limit exists.

Table 28
COMPARISON OF LIMITS FOR THE APPLICATION OF CERTAIN METALS TO AGRICULTURAL LAND IN THE U.K. VIA IRRIGATION AND SEWAGE SLUDGE DISPOSAL[21,24]

Metal	Limit (A) for metals in agricultural irrigation (mg/ℓ)	Limit (B) for annual metal addition in sewage sludge (mg/m^2)	Irrigation volume at limit (A) required to reach limit (B) per annum (ℓ/m^2)
As	0.4	33	8.3
Cu	0.5	930	186
Cr	2	3330	167
Ni	0.15	230	153
Pb	2	3330	167
Zn	1	1870	187

for cadmium, copper, nickel, lead, and zinc in sludge applied to agricultural land in the U.S. on an annual basis.[56,61] Comparing the limits for the individual metals in agricultural irrigation and sludge disposal, it is interesting to note that the sludge-disposal limits conform approximately to the "zinc-equivalent" ratio of toxicities for copper, nickel, and zinc,[24] while this apparently does not hold true for irrigation water. It is apparent that an effluent irrigation rate of 33.4 cm depth per annum containing 5 mg/ℓ of lead would impose an equivalent load in terms of total lead to the maximum allowable lead addition in sewage sludge. When it is considered that the solubility of lead in secondary effluent is typically[84] 40%, while <1% of lead in digested sludge is soluble,[85] the effects of this or a lower irrigation rate may be more severe than anticipated. Furthermore irrigation may potentially be limited by arsenic concentrations exceeding 0.1 mg/ℓ, while no national limit exists for arsenic in sludge application to agricultural land. A similar comparison applicable to the U.K. is shown in Table 28 where proposed environmental-quality standards for agricultural irrigation are compared with sludge-disposal limits. The proposed irrigation limits are based on an irrigation rate of 50 cm/annum, which would effectively apply 60 times the sludge guideline limit for arsenic and 2.7 to 3.3 times the guideline limits for copper, chromium, nickel, lead, and zinc.

REFERENCES

1. **Förstner, U. and van Lierde, J. H.**, Trace metals in water purification processes, in *Metal Pollution in the Aquatic Environment*, Förstner, U. and Wittmann, G. T. W., Eds., Springer-Verlag, New York, 1979, chap. G.
2. **Middleton, F. M. and Gromiec, M. J.**, Overview of waste water reuse practices, *Environ. Prot. Eng.*, 4, 245, 1978.
3. **Packham, R. F., Beresford, S. A. A., and Fielding, M.**, Health related studies of organic compounds in relation to re-use in the United Kingdom, *Sci. Total Environ.*, 18, 167, 1981.
4. **Funke, J. W.**, Metals in urban drainage systems and their effect on the potential reuse of purified sewage, *Water SA.*, 1, 31, 1975.
5. **Englande, A. J. and Reimers, R. S.**, Persistence of chemical pollutants in water reuse, in *Water Reuse*, Middlebrooks, E. J., Ed., Ann Arbor Science, Ann Arbor, Mich., 1982, 783.
6. **Lester, J. N.**, Significance and behavior of heavy metals in waste water treatment processes. I. Sewage treatment and effluent discharge, *Sci. Total Environ.*, 30, 1, 1983.
7. Council of the European Communities, Council directive concerning the quality required of surface water intended for the abstraction of drinking water in the member states (75/440/EEC), *Off. J. Eur. Communities*, L194, 26, 1975.
8. Council of the European Communities, Council directive relating to the quality of water intended for human consumption (80/778/EEC), *Off. J. Eur. Communities*, L229, 11, 1980.
9. **Regnier, B., Goblet, C., Genot, J., and Masschelein, W. J.**, The elimination of mineral micropollutants, *Water Sci. Technol.*, 14, 87, 1982.
10. World Health Organization, *Guidelines for Drinking Water Quality, Vol. 1, Recommendations*, World Health Organization, Geneva, 1984.
11. United States Environmental Protection Agency, National interim primary drinking water regulations, *Fed. Regist.*, 40, 59566, 1975.
12. Government of the U.S.S.R., State Committee of Standards of the U.S.S.R. Council of Ministers, All-Union State Standard 2874-73-Drinking Water, Decree No. 1972, 1973.
13. United States Environmental Protection Agency, National revised primary drinking water regulations: advance notice of proposed rulemaking, *Fed. Regist.*, 48, 45502. 1983.
14. United States Environmental Protection Agency, Water quality criteria documents; availability, *Fed. Regist.*, 45, 79318, 1980.
15. **Stoeppler, M.**, Strategies for the reliable analysis of heavy metals in man and his environment, in *Proc. Int. Conf. Heavy Metals in the Environment*, CEP Consultants Ltd., Edinburgh, 1983, 70.
16. **Commins, B. T.**, The contribution drinking water makes to man's overall exposure to inorganic substances, in *Proc. Int. Conf. Heavy Metals in the Environment*, CEP Consultants Ltd., Edinburgh, 1981, 35.
17. United States Environmental Protection Agency, Inorganic chemicals manufacturing point source category effluent limitations guidelines pretreatment standards, and new source performance standards, *Fed. Regist.*, 47, 28260, 1982.
18. Council of the European Communities, Council Directive on limit values and quality objectives for mercury discharges by the chlor-alkali electrolysis industry (82/176/EEC), *Off. J. Eur. Communities*, L81, 29, 1982.
19. Council of the European Communities, Council directive on limit values and quality objectives for cadmium discharges, (83/513/EEC). *Off. J. Eur. Communities*, L291, 1, 1983.
20. United States Environmental Protection Agency, Inorganic chemicals manufacturing point source category; effluent limitations guidelines pretreatment standards, and new source performance standards, *Fed. Regist.*, 48, 49408, 1983.
21. **Mance, G. and O'Donnell, A. R.**, Application of the European communities directive on dangerous substances to list II substances in the U.K., *Water Sci. Technol.*, 16, 159, 1983.
22. Agricultural Development and Advisory Service, Permissible Levels of Toxic Metals in Sewage used on Agricultural Land, Advisory Paper No. 10, Her Majesty's Stationery Office, London, 1971.
23. Department of the Environment and National Water Council, Report of the Working Party on the Disposal of Sewage Sludge to Land, Standing Technical Committee Report No. 5, Water Authorities Association, London, 1977.

24. Department of the Environment and National Water Council, Report of the Sub-Committee on the Disposal of Sewage Sludge to Land, Standing Technical Committee Report No. 20, Water Authorities Association, London, 1981.
25. Commission of the European Communities, Proposal for a council directive on the use of sewage sludge in agriculture, *Off. J. Eur. Community*, C264, 3, 1982.
26. Dumping at Sea Act 1974, Her Majesty's Stationery Office, London, 1974.
27. **Portmann, J. E. and Norton, M. G.**, Disposal to sea of toxic materials, *Chem. Ind.*, p.285, 1981.
28. **Bogert, I. L. and Sokol, L. J.**, Ocean dumping of sludge vs. the alternatives, *Prog. Water Technol.*, 14, 1, 1982.
29. World Health Organization, *European Standards for Drinking Water*, 2nd ed., World Health Organization, Geneva, 1970.
30. World Health Organization, *International Standards for Drinking Water*, 3rd ed., World Health Organization, Geneva, 1971.
31. **Gorchev. H. G. and Ozolins, G.**, WHO guidelines for drinking water quality, presented at Int. Water Supply Assoc. Conf., Zurich, September 6 to 10, 1982.
32. Council of the European Communities, Council directive on pollution caused by certain dangerous substances discharged into the aquatic environment of the community (76/464/EEC), *Off. J. Eur. Communities*, L129, 23, 1976.
33. Council of the European Communities, Council directive on the protection of groundwater against pollution caused by certain dangerous substances (80/68/EEC), *Off. J. Eur. Communities*, L20, 43, 1980.
34. **Mance, G., Musselwhite, C., and Brown, V. M.**, *Proposed Environmental Quality Standards for List II Substances in Water: Arsenic*, Water Research Centre, Tech. Rep. 212, Medmenham, U.K., 1984.
35. **Mance, G., Brown, V. M., Gardiner, J., and Yates, J.**, *Proposed Environmental Quality Standards for List II Substances in Water: Chromium*, Water Research Centre, Tech. Rep. 207, Medmenham, U.K., 1984.
36. **Mance, G., Brown, V. M., and Yates, J.**, *Proposed Environmental Quality Standards for List II Substances in Water: Copper*, Water Research Centre, Tech. Rep. 210, Medmenham, U. K., 1984.
37. **Brown, V. M., Gardiner, J., and Yates, J.**, *Proposed Environmental Quality Standards for List II Substances in Water: Inorganic Lead*, Water Research Centre, Tech. Rep. 208, Medmenham, U.K., 1984.
38. **Mance, G. and Yates, J.**, *Proposed Environmental Quality Standards for List II Substances in Water: Nickel*, Water Research Centre, Tech. Rep. 211, Medmenham, U.K., 1984.
39. **Mance, G. and Yates, J.**, *Proposed Environmental Quality Standards for List II Substances in Water: Zinc*, Water Research Centre, Tech. Rep. 209, Medmenham, U.K., 1984.
40. Council of the European Communities, Council directive on quality of fresh waters needing protection or improvement in order to support fish life (78/659/EEC), *Off. J. Eur. Communities*, L222, 1, 1978.
41. Agricultural Development and Advisory Service, Water Quality for Crop Irrigation: Guidelines on Chemical Criteria, ADAS leaflet No. 776, Her Majesty's Stationery Office, London, 1981.
42. The Lead in Food Regulations, Statutory Instrument No. 1254, Her Majesty's Stationery Office, London, 1979.
43. Department of the Environment, Environmental Standards. A Description of United Kingdom Practice, Pollution Paper No. 11, Her Majesty's Stationery Office, H.M.S.O., London, 1977.
44. Department of the Environment, Central Unit on Environmental Pollution, Pollution Control in Great Britain: How It Works, Pollution Paper No. 9, Her Majesty's Stationery Office, London, 1976.
45. **Otter, R. J.**, Environmental quality objectives, *Chem. Ind.*, p.302, 1979.
46. **Price, D. R. H. and Pearson, M. J.**, The derivation of quality conditions for effluents discharged to freshwaters, *Water Pollut. Control*, 78, 118, 1979.
47. **Gardiner, J.**, Monitoring and control of discharges to tidal waters, *Chem. Ind.*, p. 307, 1980.
48. **Hammerton, D., Newton, A. J., and Allcock, R.**, Determination of marine consent conditions, *Effluent Water Treat. J.*, 26, 261, 1980.
49. **Gardiner, J. and Mance, G.**, *Proposed Environmental Quality Standards for List II Substances in Water: Introduction*, Water Research Centre, Tech. Rep. 206, Medmenham, U.K., 1984.
50. United States Environmental Protection Agency, National secondary drinking water regulations, *Fed. Regist.*, 4, 42195, 1979.
51. **Lee, G. F., Jones, R. A., and Newbry, B. W.**, Water quality standards and water quality, *J. Water Pollut. Control Fed.*, 54, 1131, 1982.
52. **Train, R. E.**, *Quality Criteria for Water*, International Ideas, Philadelphia, 1980.
53. United States Environmental Protection Agency, Identification of conventional pollutants, *Fed. Regist.*, 44, 44501, 1979.
54. United States Environmental Protection Agency, Water quality criteria: request for comments, *Fed. Regist.*, 49, 4551, 1984.
55. United States Environmental Protection Agency, General pretreatment regulations for existing and new sources; postponement of effective date, *Fed. Regist.*, 47, 4518, 1982.

56. **Culp, G., Wesner, G., Williams, R., and Hughes, M. V., Jr.,** Wastewater Reuse and Recycling Technology, Noyes Data Corporation, Park Ridge, New Jersey, 1980, 1.
57. **Brown, K. W., Thomas, J. C., and Slowey, J. F.,** The movement of metals applied to soils in sewage effluent, *Water Air Soil Pollut.*, 19, 43, 1983.
58. **Cherkinskiy, S. N.,** Special features of the new Soviet standard for quality of drinking water (GOST 2874-73), *Vodosnabzh. Sanit. Tekh.*, 7, 5, 1974.
59. Government of the U.S.S.R., Ministry of Health, Maximum Permissible Concentrations of Harmful Substances in the Water of Water Courses used for Hygienic and Domestic Purposes, Report 847-70, 1970.
60. **Stöfen, D.,** The maximum permissible concentrations in the U.S.S.R. for harmful substances in drinking water, *Toxicology*, 1, 187, 1973.
61. United States Environmental Protection Agency, Municipal Sludge Management: Environmental Factors, EPA/430/9-77-004, Technical Bulletin MCD 28 United States Environmental Protection Agency, Washington, D.C., 1977, 1.
62. **Moller, U.,** The new sewage sludge ordinance and its consequences for the agricultural utilization of sewage sludge in Germany, *Korrespondenzbriefe Abwasser*, 29, 252, 1982.
63. **Davis, R. D. and Beckett, P. H. T.,** The use of young plants to detect metal accumulations in soils, *Water Pollut. Control*, 77, 193, 1978.
64. **Matthews, P. G.,** Agricultural utilisation of sludge-environmental quality objectives, in *Utilization of Sewage Sludge on Land*, Water Research Centre Conference, Water Research Centre, Medmenham, U.K., 1979, 418.
65. **Beckett, P. H. T. and Davis, R. D.,** Heavy metals in sludge — are their toxic effects additive?, *Water Pollut. Control*, 81, 112, 1982.
66. **Doyle, P. J., Lester, J. N., and Perry, R.,** Survey of literature and experience on the disposal of sewage sludge on land, Report to the Department of the Environment, DGR/480/60, Department of the Environment, London, 1978.
67. **Thormann, A.,** Utilization of sewage sludge on land. The situation, problems and development trends in Germany, in *Utilization of Sewage Sludge on Land*, Water Research Centre Conference Water Research Centre, Medmenham, U.K., 1979, 290.
68. **Scheltinga, H. M. J.,** Utilization of sewage sludge on land in Holland, in *Utilization of Sewage Sludge on Land*, Water Research Centre Conference, Water Research Centre Medmenham, U.K., 1979, 309.
69. **Hansen, J. A. and Tjell, J. C.,** Guidelines and sludge utilization practice in Scandinavia, in *Utilization of Sewage Sludge on Land*, Water Research Centre Conference, Water Research Centre, Medmenham, U.K., 1979, 317.
70. **Lester, J. N., Sterritt, R. M., and Kirk, P.W. W.,** Significance and behaviour of heavy metals in waste water treatment processes II. Sludge treatment and disposal, *Sci. Total Environ.*, 30, 45, 1983.
71. **Sterritt, R. M. and Lester, J. N.,** Concentrations of heavy metals in forty sewage sludges in England, *Water Air Soil Pollut.*, 14, 125, 1981.
72. **Johnston, A. E. and Wedderburn, R. W. M.,** The Woburn market garden experiment, 1942—1969. I, in Rothampstead Experimental Station Annual Report, Part II, Rothampstead, U.K., 1975.
73. **Dowdy, R. H. and Larson, W. E.,** The availability of sludge-borne metals to various vegetable crops, *J. Environ. Qual.*, 4, 229, 1975.
74. **Meixner, G.,** Heavy metals in sewage sludge, *Korrespondenzbriefe Abwasser*, 29, 260, 1982.
75. **Fish, H.,** Sea disposal of sludges: the U.K. experience, *Water Sci. Technol.*, 15, 77, 1983.
76. **Guarino, C. F. and Townsend, S.,** Ocean disposal in U.S. coastal waters, Philadelphia's experience and U.S. legislation, *Water Sci. Technol.*, 15, 89, 1983.
77. Convention for the Prevention of Marine Pollution by Dumping from Ships and Aircraft, Oslo, February 15, 1972, Cmnd 4984, Her Majesty's Stationery Office, London, 1972.
78. Final Act of the Intergovernmental Conference on the Convention on the Dumping of Wastes at Sea, London, November 13, 1972, Cmnd 5169, Her Majesty's Stationery Office, London, 1972.
79. **Cockburn, A. G.,** Technical control of sewage sludge disposal to sea, *Water Sci. Technol.*, 14, 17, 1982.
80. **Dakers, J. L. and King, R. P.,** Control of toxic metals of sewage treatment works, *Prog. Water Technol.*, 14, 53, 1982.
81. Department of the Environment, Sewage Sludge Disposal Data and Reviews of Disposal to Sea, Standing Technical Committee Report No. 8, Water Authorities Association, London, 1978.
82. Convention for the Prevention of Pollution from Land-Based Sources, Paris, June 4, 1974, Cmnd 5803, Her Majesty's Stationery Office, London, 1974.
83. Control of Pollution Act 1974 Her Majesty's Stationery Office, London, 1974.
84. **Oliver, B. G. and Cosgrove, E. G.,** The efficiency of heavy metal removal by a conventional activated sludge treatment plant, *Water Res.*, 8, 869, 1974.
85. **Patterson, J. W. and Hao, S. S.,** Heavy metals interactions in the anaerobic digestion system, in *Proc. 30th Ind. Waste Conf., Purdue University*, Ann Arbor Science, Ann Arbor, Mich., 1979, 544.

Chapter 4

DETERMINATION OF HEAVY METALS IN WASTEWATER MATRICES

R. M. Sterritt

TABLE OF CONTENTS

I.	Introduction	106
II.	Sampling	106
III.	Atomic Absorption Spectrophotometry	107
	A. Flame Atomization	107
	1. Sample Pretreatment	107
	a. Acid Digestion	107
	b. Dry Ashing	111
	2. Analytical Conditions	112
	B. Electrothermal Atomization	112
	1. Sample Pretreatment	112
	2. Analytical Conditions	115
	a. Instrumental Parameters	115
	b. Sample Injection	115
	c. Temperature Programs	116
	C. Mercury Determination	116
	D. Hydride Generation	118
	E. Precision Data	118
IV.	Other Methods	119
References		121

I. INTRODUCTION

It is inevitable that as environmental legislation becomes more prolific (see Chapter 3), and practices such as the disposal of sewage sludge to land become the subject of more stringent control, careful surveillance of the metallic content of sludge and other wastewater matrices will require an increasing number of analyses. It is desirable also that the analytical methodology be rapid, sensitive, inexpensive, and precise. In addition to having these desirable characteristics, the methodology must also be capable of handling the type of sample matrices under study. In the case of wastewater samples, these matrices may be extremely complex and difficult to manipulate. In this respect, the pretreatments required to convert the samples into a suitable form for analysis must be accorded an importance equal to or greater than the measuring system itself.

Of the variety of analytical techniques available for the determination of heavy metals, atomic absorption spectrophotometry fulfills most of the requirements of sensitivity, specificity, accuracy, practicability, and economy and is probably the most frequently used method for environmental samples.[1] Atomization is normally achieved either by aspiration of the liquid sample into an air-acetylene or nitrous oxide-acetylene flame[2,3] which can generate temperatures of up to about 2300 and 3000°C, respectively. Electrothermal atomization, using graphite cups, rods, or tubes has, in the last 25 years, developed to the extent where it offers an attractive alternative analytical capability for atomic absorption.[4]

The extensive proliferation of analytical methodology, particularly that based on electrothermal atomization, is evidenced by the extensive review of Slavin and Manning.[5] It must be stated, however, that only a small proportion of the current research in atomic absorption spectrophotometry and other methods, including neutron activation analysis, X-ray fluorescence, inductively coupled plasmas, and polarographic techniques, is directed towards the analysis of municipal wastewaters and similar matrices. In view of the nature of the matrices involved, these wastewaters should be afforded special attention. This chapter will be confined almost entirely to these matrices. Some of the elements considered may not qualify as heavy metals (see Chapter 1), but are included to illustrate certain principles.

II. SAMPLING

Many analysts consider sampling and sample storage to be the most crucial step in an analysis,[6] an unrepresentative sampling procedure cannot be compensated for by even the most powerful analytical technique. The problems of obtaining representative wastewater samples are generally twofold. The first lies in the removal from a large volume of material a sample which is representative of the whole. The second lies in taking a sample or samples from a process stream whose composition is representative of the average composition of the stream over a given period of time.

Sampling problems are acknowledged in U.S.[7] and U.K.[8] standard methods. The former recommends the analyst to "use great care in sampling wastewater sludges . . . No definite procedure can be given, but take every possible precaution to obtain a representative sample or one conforming to a sampling program."[7]

Representative sampling of process streams can usually be achieved by locating sampling points in regions of turbulence,[9] such as penstock flow controllers, or in rapid flows through narrow channels, provided that the velocity is high enough to maintain all the particles in suspension.[8] The sampling of sludges is more difficult. Primary sludges may be sampled during routine desludging operations.[9] A knowledge of normal practice at the works can be a distinct advantage. Where the analysis is to be done for a specific purpose — for example, regulation of the quantities of sludge applied to land — representative sampling of each load tankered away for disposal may be considered appropriate. One of the major problems involved in sampling and manipulating sludges is the fairly high suspended-solids concentration. It is advantageous if fairly large sample volumes can be taken, and even more so if these can be diluted at an early stage.[9-11] Subsampling is made a little easier by predilution; in some cases it is possible to use a wide-bore pipette with the tapered end cut off. The U.K. standard method for the determination of heavy metals in sewage sludge[12] recommends the use of a 14-mm diameter glass tube calibrated to 10 mℓ and cut off at the 10-mℓ mark. This can be held using tongs and plunged into a well-shaken bulk sample in order to obtain an aliquot which can be easily and quantitatively handled.

The design of a representative sampling program is dependent on local conditions to a great extent. Metal input may not be a continuous process, but consist of slugs lasting for a discrete period. Moreover, these slugs may not occur regularly. Where heavy metals are present in wastewater as a result of industrial discharges, up to 15-fold changes in concentration in 1 hr or less have been recorded in treatment-plant influents.[13]

Samples should be collected in borosilicate glass, or polyethylene or polypropylene containers previously leached with 10% (v/v) nitric acid for 24 hr. Containers for sewage and sludge collection may require more rigorous cleaning. A 5% (v/v) solution of Decon® 90 detergent containing 1% (v/v) disodium ethylenediaminetetraacetate[14] can be used for soaking for 24 hr prior to acid leaching.

It is normal for aqueous samples to be preserved by the addition of 1.5 mℓ of nitric acid to a liter of sample to bring the pH down to less than 2 (0.5% (v/v) nitric acid has a pH of about 1.6). For sewages and sludges this may be inadequate; a final nitric acid concentration of 1% (v/v) has been recommended.[11,15,16] If preserved in this way, metal concentrations in mixed primary sludge were unchanged for up to 6 months.[16] Caution is advised when mercury analyses are to be attempted since losses may occur, due to adsorption, after only 2 to 4 weeks.[7] The maintenance of oxidizing conditions during sample storage may help to reduce the losses of trace amounts of mercury.[17] Where flame atomic absorption is to be used for the determinations, hydrochloric acid may be preferable as a preservative, although it should not be used if silver or thallium are to be determined.[7]

III. ATOMIC ABSORPTION SPECTROPHOTOMETRY

A. Flame Atomization
1. Sample Pretreatment
a. Acid Digestion

A variety of mineral acids and acid mixtures has been used for the pretreatment of wastewater samples. The complexity of the procedures used ranges from a simple, single-stage digestion with nitric acid[12] for the determination of cadmium, chromium, copper, nickel, lead, and zinc in sludge, to a multistage digestion using nitric, perchloric, and hydrofluoric acids,[18] which brings about a more extensive mineralization of the matrix and solubilizes the more refractory metals from the insoluble inorganic matrix.[19] Hydrochloric acid is commonly used for sludge and sewage digestions; sulfuric acid has also been recommended,[20] and has been used in conjunction with nitric acid and hydrogen peroxide prior to the determination of the hydride-forming elements.[21] The general characteristics of the mineral acids for the destruction of organic matter have been described by Gorsuch.[22]

Several acids are unsuitable for the determination of specific metals. Sulfuric acid causes the formation of insoluble sulfates leading to poor recoveries of lead in particular.[23] Incomplete recoveries of bismuth as a result of coprecipitation in the presence of calcium during sulfuric acid digestions may also occur.[24] Although a procedure for the recovery of insoluble lead sulfate has been described,[25] its use represents an unnecessary additional step in the analysis and it does not recover insoluble sulfates of other metals. Perchloric acid may be unsuitable where the determination of chromium is to be undertaken[22] because of the risk of its loss as volatile chromyl chloride. It is also particularly prone to explosive reactions with organic matter, a frequent occurrence when digesting sewage sludge,[16] even when predigestion with nitric acid is carried out.

A number of reports of acid digestions being unsuitable for certain metals have been made, mainly on the basis of poor performance in comparison with alternative pretreatment methods. Thus, Moriyama et al.[26] reported better recoveries of cadmium, copper, lead, and nickel from sewage sludge by a hydrochloric acid-nitric acid digestion than by a nitric acid-perchloric acid digestion, which is an unexpected result. Poor recoveries of silver in nitric acid-hydrogen peroxide digestates have been reported by Delfino and Enderson.[27] Low recoveries of cobalt in sulphuric acid-nitric acid digestates and of tin in nitric acid-hydrogen peroxide digestates have been observed.[28] Kempton et al.[21] found poor recoveries of bismuth using nitric acid-perchloric acid digestions.

In many cases, low recoveries have been observed in situations where the digestion in question appeared to be suitable for the recovery of other metals in the same sample. This would tend to suggest that different metals require different degrees of "completeness" of the digestion in order to become fully solubilized. It is probable that very few digestions of sewage sludge, for example, are absolutely "complete" (i.e., total destruction of the organic matrix, dissolution of all inorganic forms of the metals, and zero losses due to volatilization), although this condition is probably approached most closely by the nitric acid-perchloric acid-hydrochloric acid digestion.[21] In the majority of cases, it will be necessary to be able to judge when a digestion is "complete" or to employ a standardized and reproducible technique. The former requires a considerable degree of skill and experience to achieve reproducible recoveries. It has, at least, been acknowledged that sewage sludge is a "difficult" matrix, and must be treated accordingly; thus, May and Stoeppler[29] digested sewage sludge with a nitric acid-perchloric acid mixture for 8 hr, compared to 3 hr for some foodstuffs and only 1.25 hr for certain biological materials.

Stoveland et al.[30] conducted an experiment in which aspects of both the skill component and the reproducibility component were examined. Two operators were responsible for the sulfuric acid-nitric acid digestion of three subsamples each of a batch of primary sludge. Both used the same fume cupboard, the same hotplate, and the same reagents. The digestions were as far as possible conducted in an identical manner, and at their termination could not be distinguished on a visual basis. The determination of the heavy metals in all six digestates was undertaken by only one of the operators. The results obtained are shown in Table 1. The recoveries of cadmium, copper, and zinc showed good agreement for both sets of digestions. However, one operator obtained very poor recoveries of chromium nickel and lead; these recoveries were also poor in comparison with the metal concentrations determined by a flameless technique, which was included for reference. To all intents and purposes, the data for cadmium, copper, and zinc suggested that the digestions were "complete"; moreover, they indicate that the six subsamples were obtained representatively. Therefore, it must be assumed that the digestions were incomplete in respect of the solubilization of chromium, lead, and nickel from the matrix. Effects of this type may be more likely to occur where a single fixed addition of acid to the sample is specified, which may prove insufficient for complete recovery. It is interesting to note that the metals which were inefficiently recovered by Stoveland et al.[30] were also strongly influenced by the volumes of acids actually used for digestion.[16]

A strong indication that some metals are present in a fairly refractory form can be obtained from the data of Carrondo et al.,[19] shown in Table 2. It is clear that in the case of aluminium and magnesium, only the nitric acid-perchloric acid-hydrofluoric acid digestion could achieve complete recoveries. Insofar as iron, molybdenum, and calcium were concerned, however, less powerful digestions were equally effective. For sewage effluent samples,[31] the use of perchloric and hydrofluoric acids was not required for complete recovery of aluminum and magnesium. Neither did they appear to be required for raw sewage.[31]

Table 1
THE INFLUENCE OF OPERATOR VARIATION ON THE RECOVERIES OF HEAVY METALS FROM NITRIC ACID-SULFURIC ACID DIGESTIONS OF SEWAGE SLUDGE[30]

Metal	Concentration (mg/ℓ)			Recovery by Analyst 1 as a percentage of recovery by Analyst 2
	Analyst 1	Analyst 2	Electrothermal[a]	
Cd	0.30	0.32	0.28	94
Cr	1.00	1.73	1.72	58
Cu	6.88	6.87	6.38	101
Ni	1.08	1.75	1.80	64
Pb	3.66	4.78	4.98	78
Zn	18.2	20.1	18.7	94

[a] Using analytical conditions given in Table 6.

Table 2
COMPARISON OF METHODS FOR THE RECOVERY OF ALUMINUM, CALCIUM, IRON, AND MAGNESIUM FROM SEWAGE SLUDGE

Metal	Pretreatment method	Recovery (% of maximum)
Al	H_2SO_4-HNO_3	63[a]
	HNO_3-H_2O_2	54[a]
	HNO_3-$HClO_4$-HF	100
	Ashing (450°C)	48[a]
Ca	H_2SO_4-HNO_3	89[a]
	HNO_3-H_2O_2	98
	HNO_3-$HClO_4$-HF	97
	Ashing (450°C)	100
Fe	H_2SO_4-HNO_3	100
	HNO_3-H_2O_2	97
	HNO_3-$HClO_4$-HF	100
	Ashing (450°C)	96
Mg	H_2SO_4-HNO_3	86[a]
	HNO_3-H_2O_2	86[a]
	HNO_3-$HClO_4$-HF	100
	Ashing (450°C)	81[a]

[a] Statistically different at the 0.05 probability level.

Although different workers may have apparently used the same techniques, in fact they often differ significantly with respect to the experimental details. Thus Katz et al.[32] used a bomb to digest 200 mg of dry sludge solids with 2.5 mℓ of nitric acid at 100°C overnight. Stoveland et al.[30] digested 2.0 mℓ of wet sludge (equivalent to 65 mg of dry solids) with 2.5 mℓ of nitric acid in a bomb at 130°C for 4 hr. In both cases, the technique was considered acceptable, which suggests that in the latter case, conditions were possibly more stringent than necessary. A simple nitric acid digestion was found to be acceptable by Christensen et al.[33] whereas nitric acid alone was found to be less effective by between 5 and 13% for several metals than in combination with hydrogen peroxide which in turn was of comparable efficiency to the bomb digestions in two separate comparisons.[30,32] In a study of cadmium and lead determinations in primary and digested sludge,[34] recoveries of these metals in nitric acid digestates were between 11 and 16% lower than those obtained by bomb digestion. When used in conjunction with hydrogen peroxide, however, recoveries increased to between 3 and 7% lower than those obtained with the bombs. Nitric acid in conjunction with hydrogen peroxide is generally favored by many workers,[35] although in an interlaboratory comparison[36] of sludge analysis, a pronounced negative bias was observed for nickel and zinc in nitric acid-hydrogen peroxide digestions. Rees and Hilton[37] have contended that the necessity for destroying the organic matrix is not always evident from the literature and suggested the use of a 2-hr extraction of metals from sewage sludge on a heated water bath using hydrochloric acid and hydrogen peroxide followed by conventional flame atomic absorption. The metals considered were cadmium, chromium, copper, nickel, lead, and zinc. Percentage recoveries from spiked pressed sludge cake varied between 97 and 121%. Fairly simple extraction techniques are, perhaps, justifiable since the fraction remaining in the insoluble matrix after digestion would probably not have the same potential environmental impact as the extractable fraction[38] (see Chapter 5). However, the results of Ritter et al.[39] clearly suggested that nitric acid extraction is inferior to ashing and digestion techniques for the recovery of total metals.

Aqua regia is perhaps not as popular a reagent for digestion purposes, although it has been included in two comparative studies. In comparison with five other pretreatments, which included refluxing with nitric acid for 6 hr, *aqua regia* was found to give acceptable recoveries[40] for chromium, copper, lead, manganese, nickel, and zinc in sludge. In a comparison with three other methods,[33] *aqua regia* was found to yield poor recoveries of copper, cobalt, cadmium, and chromium, although it was the best of the four pretreatments for the determination of lead. These findings were in marked contrast to those of a related study[41] of the analysis of municipal compost (incorporating sewage sludge) where the average efficiency of recovery in *aqua regia* digestions for the heavy metals studied was 95%. Although in most of the comparative work attention is focused on the efficiency of the pretreatment method, where small differences are concerned, or where reported inefficiencies are not obtained on a consistent basis, the precision of the analytical method (Section III.E), or the possibility of analytical interferences should not be overlooked as possible sources of error. These possibilities can be examined further by applying two different analytical techniques to each pretreatment technique.

Sewage and effluent samples also require careful attention in the selection of appropriate pretreatments. In general, these samples by virtue of their lower solids contents, may be expected to require smaller volumes of acids and less rigorous pretreatments than sludge samples. In the U.K., a distinction has been made between sludges[12] and sewage effluents,[42,43] the former requiring a nitric acid digestion and the latter pretreatment by "simmering" in hydrochloric acid at a final concentration of 20% (v/v) for chromium determinations[42] or evaporation with nitric acid followed by heating with hydrochloric acid for calcium determinations.[43] In the U.S.,[7] however, four different digestion techniques are described, the choice of which depends on the quantities of organic matter present in the sample, and whether this is difficult to oxidize. The exact quantities of acids to be used are not specified, with the end point of the digestion being the point at which a pale-colored, clear solution is obtained. The subjectivity inherent in this approach may contribute to the relatively poor precision for flame atomic absorption spectrophotometric analysis of wastewaters. According to Thompson,[17] it is common practice to boil acidified (1%) sewage and effluent samples for 3 min before analyzing directly.

Carrondo et al.[44] used both nitric acid-hydrogen peroxide and sulfuric acid-nitric acid digestions for sewage and effluent samples. The quantities of acids required were much less than those required for sludges,[30] although the digestion took 3 to 5 hr for completion. Both digestions appeared to be suitable for these samples, although a slightly low recovery of lead from the sulfuric acid digestion was obtained, as expected.

The time required for acid digestions is a particular disadvantage in routine analyses, especially where the attention of an operator is required. Automated techniques[45,46] can minimize this, as can the use of sealed "bombs" which permit digestion under pressure. Dramatic reductions in pretreatment time may be possible through the use of microwaves[47,48] to accelerate acid digestions to completion in 5 min or less.

b. Dry Ashing

The destruction of organic matter can be achieved in a furnace at temperatures between 300 and 800°C, depending on the metals of interest. The optimum ashing temperature is the highest at which no losses of the metal of interest occur due to volatility. Where several metals are of interest, therefore, the ashing temperature may be suboptimum in a number of cases. It is not surprising that poor recoveries of some metals as a result of their volatility at low temperatures is a common problem in the analysis of wastewater samples.[35]

In most studies on sewage sludges temperatures of 450 to 550°C have been used; however, even the lower end of this range is definitely unsuitable where cadmium is the metal of interest.[23] An ethanolic solution of magnesium nitrate can be used as an ashing aid,[49] although recoveries of cadmium from sewage sludge were lower in the presence of the aid than in its absence.[23] The only viable alternative would appear to be ashing at a lower temperature. However even at 300°C, losses of cadmium can occur.[23]

Another metal whose volatility may pose problems is lead. Although nearly complete recovery of lead, but not cadmium, was obtained by Carrondo et al.,[23] more or less similar recoveries of the two metals were obtained by Jenniss et al.[34] using the same ashing temperature. Better recoveries were obtained from digested sludge (88%) than primary sludge (80%), but whether this difference is significant is uncertain. Perfectly acceptable recoveries of cadmium from sewage sludge ashed at 550°C have been reported;[39] however, the ashing time was only 2.5 hr compared to 14 hr used by other workers.[23]

Smith[40] has reported low values for the recovery of several metals from sewage sludge ashed at 500°C; although such losses would normally be attributed to volatility, there is the possibility that inefficient extraction of the ashed residue may occur. This, in turn, could depend on two factors; namely, the inefficiency of the extractant itself or the inefficiency of the ashing stage in converting the metals into extractable forms. Smith[40] used an ashing time of 2 hr compared with the 14 hr or more of Lester and co-workers.[21,23,28] Since ashing has been reported to be a generally efficient pretreatment method,[50] the incompleteness of the ashing stage may be an important factor. Cadmium, chromium, and copper have been reported to be inefficiently recovered from dry ashing of sludge at 450°C for only 1 hr,[33] although the ashing of the residue obtained after digestion of the sludge with nitric acid and treatment with sulfuric acid was reported to be an excellent pretreatment method, which supports the conclusion that in many cases losses cannot be attributed to volatility. Kempton et al.[21] found less than 10% recovery of thallium from a sewage sludge sample ashed at 450°C for 14 hr, although for electrothermal atomic absorption (see Section III.B), an ashing temperature of 700°C was found to be suitable. This raises the possibility of a slow rate of loss over a long period. The avoidance of losses due to volatility or incomplete extraction may be possible through the use of a low-temperature oxygen plasma ashing technique.[51] This has successfully been applied to the recovery of cadmium, lead, and thallium from sewage sludge.

2. Analytical Conditions

Analytical conditions are more or less standardized for the majority of elements. Most standard methods advise the operator to follow the manufacturer's instructions. In general, the conditions shown in Table 3 will be suitable in most applications. Usually, the performance characteristics of atomic absorption are given in terms of sensitivity and detection limit. In Table 4 normal working ranges are given as an alternative which represent linear or nearly linear calibrations with acceptable baseline noise levels and which are applicable to typical concentrations of heavy metals in wastewater samples (except where there is insufficient sensitivity).

The major aspects of flame atomic absorption requiring attention[17] are the reproduction of the sample matrix in the standards, suppression of ionization where required, the addition of releasing agents, and the use of background correction.

B. Electrothermal Atomization

The much improved sensitivity of electrothermal atomic absorption is clearly of advantage for those metals present at low concentrations. It may be preferable to employ direct determination using a more sensitive analytical technique than to use preconcentration techniques for flame atomic absorption. In their comparative study, Delfino and Enderson[27] used flame atomic absorption for the determination of the major cations present, but used electrothermal atomization for silver, cadmium, chromium, copper, nickel, and lead. Although the reasons for using electrothermal atomization were not stated, it may reasonably be assumed that it was because relatively low concentrations of the latter group of metals were present.

In addition to its sensitivity, the other major attraction of electrothermal atomization is that the direct injection of some samples may be employed. Since the analytical technique itself is time consuming, the avoidance of lengthy pretreatments seems attractive.

1. Sample Pretreatment

Relatively "clean" samples (for example, effluents which have been filtered[50]) may not require pretreatment at all prior to electrothermal atomic absorption. It is probably not surprising that a number of workers have examined the possibility that successful analyses could be undertaken even if some suspended solids were to be introduced into the atomizer. Kunselman and Huff[52] stated that the technique "has the capability to determine total metals without chemical pretreatment, provided the aqueous phase can be sampled in a reproducible manner". However, electrothermal atomization has found application for wastewater samples where predigestion techniques have been used, particularly in the analysis of selenium. A predominant reason why this element has received so much attention is that flame atomization suffers from high background noise levels due to absorption of the primary resonance line energy by the flame itself. Nitric acid-perchloric acid[53] and nitric acid-hydrogen peroxide[54,55] digestions have been used, although Henn[54] acknowledged the fact that the acid digestion was a major cause of low recoveries of the element.

Selenium determinations by electrothermal atomic absorption appear to benefit from the addition of nickel nitrate[55] (0.1 to 5.0%, depending on the solids concentration in the sample), or molybdenum[54] to improve stability, lessen interferences, and reduce volatility during the ashing stage. Furthermore, the susceptibility to interferences from inorganic components of the matrix can be avoided by passing the digestate through ion-exchange resins prior to determination of selenium.

Attempts to determine selenium directly in suspended solids-containing samples (raw and settled sewage) without predigestion have met with failure,[53] with recoveries of 27% or less being obtained. This, however, was in all probability due to poor sampling of the solids since a 5 µg syringe with platinum needle was used; no real attempt to sample the matrix

Table 3
ANALYTICAL CONDITIONS FOR THE FLAME ATOMIC ABSORPTION SPECTROPHOTOMETRIC DETERMINATION OF HEAVY METALS IN WASTEWATER MATRICES

Metal	Slit width (nm)	Wavelength (nm)	Flame type[a]	Notes
Ag	0.7	328.1	Air-C_2H_2 (O)	
Al	0.7	309.3	N_2O-C_2H_2 (R)	Ionization suppressant
As	0.7	193.7	Air-C_2H_2 (R)	
Be	0.7	234.9	N_2O-C_2H_2 (R)	
Bi	0.2	223.0	Air-C_2H_2 (O)	
Ca	0.7	422.7	Air-C_2H_2 (O)	Ionization suppressant
Cd	0.7	228.8	Air-C_2H_2 (O)	
Co	0.2	240.7	Air-C_2H_2 (O)	
Cr	0.2	357.9	Air-C_2H_2 (R)	
Cu	0.7	324.8	Air-C_2H_2 (O)	
Fe	0.2	248.3	Air-C_2H_2 (O)	
Mg	0.7	285.2	Air-C_2H_2 (O)	Add 1% LaCl
Mn	0.2	279.5	Air-C_2H_2 (O)	
Mo	0.7	313.3	N_2O-C_2H_2 (R)	
Ni	0.2	232.0	Air-C_2H_2 (O)	
Pb	0.7	283.3	Air-C_2H_2 (O)	
Sb	0.2	217.6	Air-C_2H_2 (O)	
Sn	0.7	286.3	N_2O-C_2H_2 (R)	
Te	0.2	214.3	Air-C_2H_2 (O)	
V	0.7	318.4	N_2O-C_2H_2 (R)	Ionization suppressant
Zn	0.7	213.9	Air-C_2H_2 (O)	

[a] O, oxidizing; R, reducing.

Table 4
TYPICAL WORKING RANGES FOR FLAME ATOMIC ABSORPTION SPECTROPHOTOMETRY COMPARED WITH TYPICAL CONCENTRATIONS IN WASTEWATER SAMPLES

Metal	Working range (mg/ℓ)	Typical concentration (mg/ℓ) Primary Sludge[a]	Raw Sewage[b]
Ag	0.05—4.0	2	0.01
Al	1—50	200	3.0
As	1—50	0.5	0.01
Be	0.02—2.0	0.25	—
Bi	0.2—20	1.5	0.005
Ca	0.05—5.0	1000	100
Cd	0.005—1.0	2.5	0.01
Co	0.05—5.0	5	0.01
Cr	0.05—5.0	50	0.2
Cu	0.05—5.0	50	0.2
Fe	0.05—5.0	150	2
Mg	0.005—0.5	100	10
Mn	0.03—3.0	25	0.1
Mo	1—15	0.5	0.01
Ni	0.05—5.0	10	0.1
Pb	0.2—10.0	50	0.1
Sb	0.3—30	0.5	0.05
Sn	1—20	5	0.01
Te	0.3—30	—	—
Tl	0.5—50	—	<0.001
V	1—50	5	—
Zn	0.01—1.0	100	0.5

[a] Calculated from data in Volume II, Chapter 3.
[b] Calculated from data in Volume II, Capter 1.

reproducibly at low volume was made. Sample injection using micropipettes of the Eppendorf type, or simila designs, with interchangeable tips, appears to be suitable for the direct injection of sewage effluents. In addition to selenium, direct injection techniques have been found to be applicable to the other metalloids of concern, namely arsenic, antimony, and tellurium.[52] Flame atomic absorption is insufficiently sensitive for these elements in the majority of wastewater matrices.[21] Suppression or enhancement effects due to the matrix were effectively overcome by using the method of standard additions and recoveries of all these elements in the range 90 to 110% from sewage effluent were obtained.[52]

The principle of direct injection has been extended to the determination of other metals in sewage effluent and similar materials where the traditional flame atomic absorption techniques are not as difficult to apply or insensitive. In addition to the four metalloids discussed above, beryllium, cadmium, lead, and silver have been determined in "industrial and municipal wastes" by direct injection,[56] yielding results very similar to those obtained by flame atomic absorption. A variety of wastewater samples has also been analyzed successfully using similar techniques.[57]

Lester and co-workers have extended the principle of direct injection to include raw sewages[44] and primary and mixed primary sludges.[11,15,30] Reproducible sampling of these matrices to obtain injection volumes of <100 μℓ is highly dependent on the homogeneity of the matrix. Sufficient homogenization and dispersion of even thick sludges, having up to 3% (w/v) total solids concentrations, has been achieved using a device which mechanically disrupts large particles.[11,15] The device (Ultra-Turrax, T45N, Scientific Instrument Co.) is a homogenizer based on the stator-rotor principle, and also generates ultrasonic waves. It was necessary to replace the original stainless-steel shaft with a replica machined from titanium since the former contributed particularly high blank values for nickel and chromium.[15]

A normal procedure for the homogenization of a sewage sludge sample involves dilution of 200 to 1000 mℓ with 1% (v/v) nitric acid, followed by a further tenfold dilution in 1% (v/v) nitric acid prior to homogenization of 1000 mℓ of the resultant mixture at 6000 to 10000 rev/min for 7 to 10 min.[11] With experience, further studies[15] showed that all but the most difficult sludges could be completely homogenized at 8000 rev/min for 5 min. The suitability of the treated sample for direct injection is thus due to the combined effects of homogenization and extensive dilution of the original sample. Raw sewage and sewage effluent samples (200 mℓ) could be dispersed effectively under the same conditions for homogenization but without the necessity for prior dilution.[44]

2. Analytical Conditions
a. Instrumental Parameters

The electrothermal atomic absorption instrumental parameters (absorbing wavelength and slit width) will in general be the same as those required for flame atomic absorption. For zinc determinations, however, the nonsensitive line at 307.6 nm can be used routinely to avoid dilution of the samples. Dependent on the samples, other elements may be determined using either sensitive or nonsensitive wavelengths. The relevant data given in Table 6 correspond to the sensitive wavelengths given in Table 3. Alternative nonsensitive wavelengths can be used for the determination of aluminum (257.5 nm), calcium (239.9 nm), iron (305.9 nm), and magnesium (202.6 nm).[31] These afford sensitivities approximately 6, 150, 20, and 30 times lower than the primary absorbing wavelengths, respectively.

b. Sample Injection

The mode of introduction of the sample, which normally will involve a micropipette dispensing 20 to 100 $\mu\ell$, is clearly a crucial stage in the analysis, and the inappropriate choice of pipette may be a major cause of failure of the electrothermal technique for undigested samples.[53]

A certain degree of operator skill is required to observe the following general requirements of the technique.[11,15] Vigorous shaking of the sample prior to removal of a small aliquot is essential in order to obtain a homogenous suspension. Plastic disposable pipette tips should be rinsed in dilute nitric acid, distilled water, and sample prior to injection. A major cause of error is the incomplete ejection of the sample aliquot from tips which have lost their nonwetting properties. Frequent replacement and careful attention to the injection technique are required.

Injections should be made at least in duplicate. Many analysts believe that the method of standard additions is absolutely mandatory for electrothermal atomization; however, with experience involving a particular matrix, it may be found that a calibration curve will give acceptable results provided that the calibration is checked every 10 to 15 injections. Although the susceptibility of electrothermal atomization to interferences is frequently the reason for reservations regarding the technique, marked interference effects are not particularly common in wastewater matrices. Carrondo et al.[58] examined the effects of a range of organic and inorganic conditioners added to sewage sludge on the determination of heavy metals. The conditioners used were lime and ferrous sulphate at rates of addition of up to 37.5% (w/w dry sludge solids), lime and ferric chloride at up to 22.5% (w/w), aluminum chlorohydrate at up to 4.5% (w/w), and cationic polyelectrolytes at up to 1.5% (w/w). The only marked interferences were obtained for aluminum chlorohydrate, which depressed the analytical signal for cadmium, chromium, copper, and nickel. The other conditioners and polyelec-

Table 5
ASHING TEMPERATURES

Metal	Ashing temperatures (°C)	
	Lowest	Highest
Ag	400	850
Be	1200	1500
Bi	350	600
Cd	250	400
Pb	350	700
V	1000	1600

Note: Values given are examples of the range of ashing temperatures reported for electrothermal atomic absorption spectrophotometric analysis of wastewater samples.

trolytes did not interfere when present at levels typical of those achieved normally in sludge treatment. The use of simultaneous background correction is very important, particularly for the volatile elements. For the nonvolatile (e.g., copper and chromium) and refractory (e.g., molybdenum) metals, background correction has not always been found necessary, although this may have been fortuitous. With the advent of tungsten-halide systems (in addition to the deuterium arc), background correction can be provided at all wavelengths normally used.

c. Temperature Programs

Temperature programs, and particularly the ashing or charring stage, are the most important aspect of the analysis. Fuller[4] has given maximum recommended ashing temperatures for a comprehensive range of elements. They rarely, however, correspond to the actual temperatures used, as shown by the examples in Table 5. In some cases, the maximum and minimum ashing temperatures differ by up to twofold. As a general rule, optimum ashing temperatures should be established for each new sample matrix under study. Optimum ashing temperatures for lead may, however, vary considerably with the "age" of the graphite furnace or tube in use. This effect[16] is shown in Figure 1 and may explain in particular the wide variations in recorded ashing temperatures shown in Table 5. It is interesting to note that although furnace age markedly affected the maximum tolerable ashing temperature,[16] the sample matrix did not, despite the fact that lead ashing may be dependent on the extent to which organic species of the element are present.[59] A general indication of ashing and atomization temperatures suitable for wastewater matrices is given in Table 6. Many of these have been based on the maximum temperatures which do not cause a reduction in the analytical signal in the subsequent atomization stage. In view of the apparent temperature changes with age, it may be prudent to set ashing temperatures about 100°C lower than this maximum.[24]

C. Mercury Determination

Mercury requires special attention for several reasons. First, its high volatility precludes the use of some of the rigorous pretreatment techniques described above. Second, flame atomic absorption is far too insensitive for mercury determinations at typical concentrations in the matrices under consideration here. Third, a conventional ashing stage cannot be

Table 6
ANALYTICAL CONDITIONS FOR ELECTROTHERMAL ATOMIC ABSORPTION SPECTROPHOTOMETRIC DETERMINATIONS OF HEAVY METALS IN WASTEWATER MATRICES.

Metal	Temperature (°C) Ashing	Atomization	Working range[a] (µg/ℓ)
Ag	400	2700	5—75
Al	1200[b]	2800[c]	50—1000
Bi	600	2300	
Ca	1200[b]	2800	10—80
Cd	250	2100	5—30
Co	1100[b]	2700	50—500
Cr	1100[b]	2700	20—200
Cu	800	2800	50—400
Fe	1200[b]	2800	50—500
Mg	1200[b]	2800	2—40
Mn	1000	2700	10—100
Mo	1750[b]	2800[d]	10—75
Ni	900	2800	100—1000
Pb	450	2300	25—400
Sn	900[e]	2700	200—2000
Tl	700	2700	
V	1000	2900[f]	
Zn	500	2500	200—2000

[a] Working range quoted is for 20-µℓ injections.
[b] If ramp facility is not available, use a preliminary ashing stage of 400°C.
[c] Analysis particularly sensitive to small variations in atomization temperature.
[d] Severe memory effects encountered, even with atomization time of 8 sec.
[e] Analysis improved if long ashing times (up to 150 sec) are used. Injected sample neutralized by injection of 17% (v/v) ammonia solution.
[f] Memory effects reduced by using pyrolytically coated graphite tubes and ballistic heating during atomization.

incorporated into a temperature program for electrothermal atomization. Mercury is normally determined by the cold vapor technique[7,60] in which organic forms of the metal are reduced to the element and the atomic vapor is swept by a stream of inert gas into an argon-hydrogen or nitrogen-hydrogen flame. Because of the potential problem of volatile loss, one of the digestion techniques recommended is limited to a temperature of 95°C. It normally involves oxidation of a sample treated with sulfuric and nitric acids using potassium permanganate and potassium persulfate.[61] Excess oxidant is reduced using hydroxylamine sulfate prior to analysis. For samples containing large quantities of organic matter, oxidation at 70°C for 4 hr has been recommended.[7] In the U.K., a tentative standard method[60] for mercury in sludge involves heating to only 55°C for 16 to 24 hr with sulfuric acid, the subsequent permanganate oxidation step being conducted at ambient temperature for a further 24 hr or more. A rapid technique,[62] involving digestion with bromine for 10 min, has been described for the pretreatment of sewage effluents prior to mercury determinations.

In view of the fact that the discussion in Section I would suggest that incomplete recovery

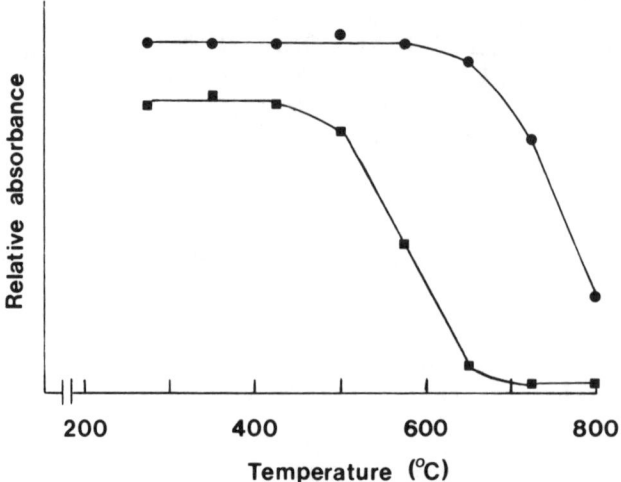

FIGURE 1. The effect of graphite tube age on ashing temperatures for lead determination; (■), old tube; (●), new tube.

of metals from complex matrices is a common problem, some of the techniques for mercury recovery may appear insufficiently rigorous. Digestion of sewage sludge by boiling nitric acid-perchloric acid mixtures[29] has proven successful for mercury determinations provided that special quartz tubes are used; to the top of each tube is connected (via a ground glass joint) a U-tube in which ultrapure water may be placed. Retention of the mercury within the digestion vessel occurs by a mixture of reflux, with the upper region of the tube acting as a condenser, and absorption of the vapor in the U-tube.[29]

D. Hydride Generation

Arsenic and selenium are normally determined by hydride generation since standard flame techniques are insensitive and subject to significant absorption of radiation by the flame itself.[63] Antimony, bismuth, and tellurium have also been determined successfully in sewage sludge by hydride generation.[21]

The hydride-forming metalloids can exist in several different valence states which may affect the analytical signal, probably because the reduction kinetics for each are different. Prereduction of the sample to ensure that all of the determinand is in the correct oxidation state may be required.[64] Antimony and arsenic may be reduced to the trivalent state using iodide,[21] while all tellurium can be converted into the tetravalent state by boiling with *aqua regia* (3:1 [v/v] hydrochloric acid/nitric acid) for 10 min.[21] Although a stannous chloride-potassium iodide mixture has been used for prereduction, it is reported to be incompatible with hydride generation using sodium borohydride.[65] For bismuth determinations, prereduction was not found to be necessary by Kempton et al.[21] since the trivalent form is virtually the only naturally occurring state.[64]

The hydride forming elements can be recovered successfully from sewage sludge using sulfuric acid-nitric acid digestions[7,20] with the addition of hydrogen peroxide[20,21] or perchloric acid.[65] Oxidizing conditions must be maintained to avoid the loss of arsenic.[65]

E. Precision Data

Precision data are probably best expressed in terms of the standard deviation of the results of a series of determinations on the same sample. However, the significance of the standard deviation figure can only be assessed by relating it to the concentration determined; hence, the relative standard deviation (RSD, the standard deviation as a percentage of the mean) is often used. This too has certain drawbacks since quite often the RSD is dependent on concentration, with greater precision being obtained at higher determinand concentrations. However, RSD will be used here preferentially because it is relevant to the concentration ranges normally found in wastewater matrices.

It has been pointed out frequently that flame atomic absorption is inherently more precise than electrothermal atomic absorption, with typical RSDs of <1% compared to 1 to 10% for the latter. This is certainly true for repeated determinations on the same solution. However, larger errors are introduced by processing samples through complex pretreatment procedures. Since in the case of flame atomization, some kind of pretreatment cannot usually be avoided, except perhaps in the case of some sewage effluents, the pretreatment must be considered an integral part of the analysis and the RSDs reported accordingly. Thus, the data given here are for *replicate samples,* each processed through the pretreatments and analyzed separately, rather than *replicate aspirations* or *injections* of the same pretreated sample (or standard solution) which are sometimes used as the basis of precision data (sometimes termed repeatability).[17]

Table 7 shows precision data for heavy metal determinations in a range of wastewater matrices. Since it is the objective of this chapter to specifically address the determination of metals in wastewater matrices, values are given for sludges, sewages, and effluents since precision may decrease with decreasing concentration of the determinand in some cases. This effect is probably responsible for the high RSDs for cadmium determinations in sewage effluent and the tin determinations in sewage sludge, which were below the normal working range for this element. The opposite trend occurs for chromium and nickel, both elements which are particularly sensitive to acid digestion conditions (see Table 1).[30] This may be due to the greater ease of dissolution of these metals from the less complex sewage and effluent matrices under digestion conditions less critical in terms of acid volumes added. A similar effect might have been expected for lead, but for the fact that flame atomization proved insufficiently sensitive for sewage and effluent samples.

Within the ranges quoted for flame atomic absorption, the lower RSDs tended to be obtained from dry ashing, presumably due to the relative simplicity of the practical technique, and from bomb digestions, presumably due to the simplicity of the technique, its tendency to attain more extensive matrix destruction, and the prevention of loss due to volatility and spattering.[39]

IV. OTHER METHODS

In addition to atomic absorption spectrophotometry, about seven other major categories of analytical techniques are available for the determination of heavy metals, including colorimetric methods, emission spectrophotometry, voltammetry, ion electrodes, X-ray fluorescence, neutron activation analysis, and mass spectrometry.[66] In the wastewater laboratory and for routine analytical purposes, the latter four are not widely applicable because of low sensitivity, high cost, or limitations in the number of elements which can be determined, or a combination of these factors.

Standard colorimetric methods are described for many heavy metals in the U.S.,[7] although the method of choice in the majority of cases, and especially where wastewater matrices are involved, is atomic absorption. In the U.K. until the publication of new analytical techniques was initiated, the recommended methods for heavy metal determinations[67] were colorimetric techniques. These were evaluated for the determination of heavy metals in sewage sludge by Stoveland et al.[30] In general, the determinations were free of error, but had poor precision, with RSDs for nickel, lead, and zinc in sewage sludge of 27, 38, and 35%, respectively.

Polarographic techniques for wastewater samples have not been widely adopted. Although

Table 7
PRECISION DATA FOR HEAVY METAL DETERMINATIONS IN WASTEWATER SAMPLES BY FLAME, ELECTROTHERMAL, AND HYDRIDE GENERATION ATOMIC ABSORPTION

	Relative Standard Deviation (%)							
	Sludge			Sewage[a]			Effluent	
Metal	F	E	H	F	E	H	F	E
Ag	22—30	5			5			
Al	4—6			5—13		—[b]	11	
As	3[c]		4[c]		3—8	11		
Be					3			
Bi	4—14	7	4—12					
Ca	2—3			4—8	5		6—7	4
Cd	4—11	4—8		15—20	3—15		15—26	15
Co	8—24	5						
Cr	3—30	6—10		1—4	1—2		2—5	2
Cu	3—14	3—6		2—7	3—6		8—14	8
Fe	3—6			5—8	6—8		10—12	10
Mg	2—3			6—8	4—5		5—10	5
Mn	1—21	9						
Mo	5—16	6—19						
Ni	2—26	5—8		2—7	2—3		7—8	3
Pb	1—17	3—9		—	3—4		—	5
Sb	8[c]		8[c]		3—4			
Se					5—15			
Sn	15—34	13						
Te	2[c]		17—22[c]		4			
Tl	11—36[d]	4—11						
V	3—8	12—15						
Zn	1—5	3—7		3—5	2—5		7—9	3

Note: F, flame; E, electrothermal; H, hydride generation.

[a] Where sample type has not been exactly specified (e.g., "wastewater effluent", "wastewater", or "sewage treatment plant samples"), data have been included in the columns corresponding to sewage.
[b] A dash indicates that the technique was applied but proven insufficiently sensitive.
[c] Samples spiked to bring concentration within working range.
[d] Method of standard additions used.

methods had been described for the determination of cadmium, copper, nickel, lead, and zinc by polarography in the 14th edition[68] of *Standard Methods for the Examination of Water and Wastewater* in the U.S., the material was subsequently omitted from the 15th since it was judged antiquated and unworthy of standard status. Polarographic determinations of cadmium, copper, nickel, and zinc have been given brief consideration by Allen and Minear,[61] and accorded the status of secondary methods. A method for the determination of cadmium, copper, lead, and zinc in water and wastewater by different pulse anodic stripping voltammetry (DPASV) has been described,[69] with reported RSDs for a sample containing concentrations of heavy metals typical of raw sewage ranging from 3% for zinc to 14% for cadmium. Good precision, with an RSD of only 2.5%, has been reported for the DPASV determinations of lead in sewage sludge following pretreatment by low-temperature ashing.[51] The method is subject to a number of interferences from other elements, although these may be eliminated. Examples of such elimination include the use of ascorbic acid as a masking agent in the presence of selenium and the reduction of interfering ferric ions to ferrous ions. The method is not applicable to a comprehensive range of elements.

The use of inductively coupled plasma (ICP) spectrophotometry is an attractive alternative to traditional atomic absorption[70] and is considered a good technique for water and wastewater analysis.[71] It offers the possibility of simultaneous analysis[72,73] and can be used in the emission mode,[73] but is subject to limitations in terms of the necessity for sample pretreatment.[72]

X-ray fluorescence spectroscopy[74] and neutron activation analysis[75] are both selective, precise, and accurate for heavy metals in dried samples (e.g., sewage sludge) and have been used to a limited extent in wastewater analyses, although the instrumentation required for the latter technique is probably beyond the scope of most routine analytical laboratories. Neutron activation analysis is probably of most use in certification of other methods. Katz et al.[32] used neutron activation analysis as a technique to check on recoveries of metals from sewage sludge by conventional digestion techniques, although the determinations of cadmium, lead, and iron were not attempted due to poor sensitivity.

Photon activation analysis has been used to determine heavy metals in sewage sludge.[76] The claimed advantages of this method are that it can be used for predominantly aqueous samples and that it permits the instrumental determination of nickel, lead, and thallium at trace levels which might be problematic in neutron activation analysis.

Another technique which qualifies as being "definitive" is mass spectrometry using isotope dilution analysis. This has been used for thallium and lead in sewage sludge[51] with good precision, although DPASV and electrothermal atomic absorption were equally effective for the latter metal and in the majority of cases would be preferable techniques on the grounds of cost and time.

REFERENCES

1. **Dulka, J. J. and Risby, R. H.**, Ultratrace metals in some environmental and biological systems, *Anal. Chem.*, 48, 640A, 1976.
2. **Coleman, R. F.**, Comparison of analytical techniques for inorganic pollutants, *Anal. Chem.*, 46, 989A, 1974.
3. **Willard, H. W., Merritt, L. L., Dean, J. A., and Settle, F. A.**, *Instrumental Methods of Analysis*, 6th ed., Wandsworth, Belmont, Calif., 1981, 127.
4. **Fuller, C. W.**, *Electrothermal Atomisation for Atomic Absorption Spectrometry*, Analytical Sciences Monograph No. 4, The Chemical Society, London, 1977.
5. **Slavin, W. and Manning, D. C.**, Graphite furnace interferences, a guide to the literature, *Prog. Anal. At. Spectrosc.*, 5, 243, 1982.
6. **Stoeppler, M.**, Strategies for the reliable analysis of heavy metals in man and his environment, in *Proc. Int. Conf. Heavy Metals in the Environment*, CEP Consultants Ltd., Edinburgh, 1983, 70.
7. American Public Health Association, American Water Works Association, Water Pollution Control Federation, *Standard Methods for the Examination of Water and Wastewater*, 15th ed., American Public Health Association, Washington, D.C., 1981, 141.
8. Standing Committee of Analysts, Department of the Environment, The Sampling and Initial Preparation of Sewage and Waterworks Sludges Soils, Sediments and Plant Material Prior to Analysis, Methods for the Examination of Waters and Associated Materials, Her Majesty's Stationery Office, London, 1977.
9. **Lester, J. N., Harrison, R. M., and Perry, R.**, The balance of metals through a sewage treatment works. I. Lead, cadmium and copper, *Sci. Total Environ.*, 12, 13, 1979.
10. **Stoveland, S., Astruc, M., Lester, J. N., and Perry, R.**, The balance of heavy metals through a sewage treatment works. II. Chromium, nickel and zinc, *Sci. Total Environ.*, 12, 25, 1979.
11. **Lester, J. N., Harrison, R. M., and Perry, R.**, Rapid flameless atomic absorption analysis of the metallic content of sewage sludges. I. Lead, cadmium and copper, *Sci. Total Environ.*, 8, 153, 1977.
12. Standing Committee of Analysts, Department of the Environment, Cadmium, Chromium, Copper, Lead, Nickel and Zinc in Sewage Sludges by Nitric Acid/AAs 1981, Methods for the Examination of Waters and Associated Materials, Her Majesty's Stationery Office, London, 1982.
13. **Oliver, B. G. and Cosgrove, E. G.**, The efficiency of heavy metal removal by a conventional activated sludge treatment plant, *Water Res.*, 8, 869, 1974.

14. **Henricksen, A. and Balmer, R.,** Sampling, preservation and storage of water samples for analysis of metals, *Vatten,* 1, 33, 1977.
15. **Stoveland, S., Astruc, M., Perry, R., and Lester, J. N.,** Rapid flameless atomic absorption analysis of the metallic content of sewage sludges. II. Chromium, nickel and zinc, *Sci. Total Environ.,* 9, 263, 1978.
16. **Carrondo, M. J. T., Lester, J. N., Perry, R., and Stoveland, S.,** Analysis of Heavy Metals in Sewage Sludge, Sewages and Final Effluent, Final Report to the Department of the Environment, Contracts DGR/480/66 and DGR/480/240, Imperial College, London, 1978.
17. **Thompson, K. C.,** Atomic Absorption Spectrometry, 1979 Version, Methods for the Examination of Waters and Associated Materials, Her Majesty's Stationery Office, London, 1979.
18. **Agemain, H. and Chau, A. S. Y.,** An atomic absorption method for the determination of 20 elements in lake sediments after acid digestion, *Anal. Chim. Acta,* 80, 61, 1975.
19. **Carrondo, M. J. T., Lester, J. N., and Perry, R.,** An investigation of a flameless atomic-absorption method for determination of aluminum, calcium, iron and magnesium in sewage sludge, *Talanta,* 26, 929, 1979.
20. **Arbab-Zavar, M. H. and Howard, A. G.,** Automated procedure for the determination of soluble arsenic using hydride generation atomic absorption spectroscopy, *Analyst (London),* 105, 744, 1980.
21. **Kempton, S., Sterritt, R. M., and Lester, J. N.,** Atomic-absorption spectrophotometric determination of antimony, arsenic, bismuth, tellurium, thallium and vanadium in sewage sludge, *Talanta,* 29, 675, 1982.
22. **Gorsuch, T. T.,** *The Destruction of Organic Matter,* Pergamon Press, New York, 1970.
23. **Carrondo, M. J. T., Perry, R., and Lester, J. N.,** Comparison of electrothermal atomic absorption spectrometry of the metal content of sewage sludge with flame atomic absorption spectrometry in conjunction with different pretreatment methods, *Anal. Chim. Acta,* 106, 309, 1979.
24. **Cantle, J. E.,** Practical techniques, in *Atomic Absorption Spectrometry, Techniques and Instrumentation in Analytical Chemistry,* Vol. 5, Cantle, J. E., Ed., Elsevier, Amsterdam, 1982, 37.
25. **Hanson, N. W.,** *Recommended Methods of Analysis,* 2nd ed., The Society for Analytical Chemistry, London, 1974.
26. **Moriyama, K., Watanabe, A., Sugiura, S., Arayashiki, H., and Mori, Y.,** Atomic absorption spectrophotometric determination of cadmium, lead, nickel and copper in sewage sludge, *Gesuido Kokaishi,* 19, 68, 1982.
27. **Delfino, J. J. and Enderson, R. E.,** Comparative study outlines methods of analysis of total metal in sludge, *Water Sewage Works,* p. R32, 1978.
28. **Sterritt, R. M. and Lester, J. N.,** Determination of silver, cobalt, manganese, molybdenum and tin in sewage sludge by a rapid electrothermal atomic-absorption spectroscopic method, *Analyst (London),* 105, 616, 1980.
29. **May, K. and Stoeppler, M.,** Pretreatment studies with biological and environmental materials. IV. Complete wet digestion in partly and completely closed quartz vessels for subsequent trace and ultratrace mercury determination, *Fresenius Z. Anal. Chem.,* 317, 248, 1984.
30. **Stoveland, S., Astruc, M., Perry, R., and Lester, J. N.,** Comparison of flameless atomic absorption for the analysis of the metallic content of sewage sludge with flame atomic absorption and colorimetric methods, *Sci. Total Environ.,* 13, 33, 1979.
31. **Carrondo, M. J. T., Lester, J. N., and Perry, R.,** Determination of aluminum, calcium, iron and magnesium in sewages and sewage effluent by a rapid electrothermal atomic absorption spectroscopic method, *Analyst (London),* 104, 831, 1979.
32. **Katz, S. A., Jennis, S. W., Mount, R., Tout, R. E., and Chatt, A.,** Comparison of sample pretreatment methods for the determination of metals in sewage sludges by flame atomic absorption spectrometry, *Int. J. Environ. Anal. Chem.,* 9, 209, 1981.
33. **Christensen, T. H., Pedersen, L. R., and Tjell, J. C.,** Comparison of four methods for digestion of sewage sludge samples for analysis of metals by atomic absorption spectrophotometry, *Int. J. Environ. Anal. Chem.,* 12, 41, 1982.
34. **Jennis, S. W., Katz, S. A., and Mount, T.,** A comparison of sample preparation methods for determination of cadmium and lead in sewage sludges by AAS, *Am. Lab.,* 12, 18, 1980.
35. **Sterritt, R. M. and Lester, J. N.,** Atomic absorption spectrophotometric analysis of the metal content of waste water samples, *Environ. Technol. Lett.,* 1, 402, 1980.
36. **Adelman, H., Jennis, S. W., and Katz, S. A.,** Interlaboratory analysis of sewage sludge, *Am. Lab.,* 13, 31, 1981.
37. **Rees, T. D. and Hilton, J.,** A rapid method for the determination of heavy metals in sewage sludges, *Lab. Prac.,* 27, 291, 1978.
38. **Stover, R. C., Sommers, L. E., and Silviera, D. J.,** Evaluation of metals in wastewater sludge, *J. Water Pollut. Control Fed.,* 48, 2165, 1976.
39. **Ritter, C. J., Bergman, S. C., Cothern, C. R., and Zamierowski, E. E.,** Comparison of sample preparation techniques for atomic absorption analysis of sewage sludge and soil, *At Absorpt. Newsl.,* 17, 70, 1978.

40. **Smith, R.,** Evaluation of various techniques for the pretreatment of sewage sludges prior to trace metal analysis by atomic absorption spectrophotometry, *Water S.A.,* 9, 31, 1983.
41. **Christensen, T. H.,** Comparison of methods for preparation of municipal compost for analysis of metals by atomic absorption spectrophotometry, *Int. J. Environ. Anal. Chem.,* 12, 211, 1982.
42. **Standing Committee of Analysts,** Department of the Environment, Chromium in Raw and Potable Waters and Sewage Effluents, Methods for the Examination of Waters and Associated Materials, Her Majesty's Stationery Office, London, 1980.
43. **Standing Committee of Analysts,** Department of the Environment, Calcium in Waters and Sewage Effluents by Atomic Absorption Spectrophotometry, Methods for the Examination of Waters and Associated Materials, Her Majesty's Stationery Office, London, 1977.
44. **Carrondo, M. J. T., Perry, R., and Lester, J. N.,** Comparison of a rapid flameless atomic absorption procedure for the analysis of the metallic content of sewages and sewage effluents with flame atomic absorption methods, *Sci. Total Environ.,* 12, 1, 1979.
45. **Budna, K. W. and Knapp, G.,** Continuous decomposition of organic materials with hydrogen peroxide-sulphuric acid, *Fresenius Z. Anal. Chem.,* 294, 122, 1979.
46. **Frank, A.,** Automated wet ashing and multi-metal determination in biological materials by atomic-absorption spectrometry, *Z. Anal. Chem.,* 279, 101, 1976.
47. **Barrett, P., Davidson, L. J., Penaro, K. W., and Copeland, T. R.,** Microwave oven-based wet digestion technique, *Anal. Chem.,* 50, 1021, 1978.
48. **West, M. H., Molina, J. F., Yaun, C. L., Davis, D. G., and Chauvin, J. V.,** Determination of metals in waters and organic materials by flameless atomic absorption spectrometry with a wire-loop atomizer, *Anal. Chem.,* 51, 2370, 1979.
49. **Friend, M. T., Smith, C. A., and Wishant, D.,** Ashing and wet oxidation procedures for the determination of some volatile trace metals in foodstuffs and biological materials by AAS, *At. Absorpt. Newsl.,* 16, 46, 1977.
50. **Guillaumin, J. C.,** Determination of trace metals in power plant effluents, *At. Absorpt. Newsl.,* 13, 135, 1974.
51. **Waidmann, E., Hilpert, K., Schladdt, J. D., and Stoeppler, M.,** Determination of cadmium, lead and thallium in materials of the Environmental Specimen Bank using mass spectrometric isotope dilution analysis (MS-IDA), *Freseius Z. Anal. Chem.,* 317, 273, 1984.
52. **Kunselman, G. C. and Huff, E. A.,** The determination of arsenic, antimony, selenium, and tellurium in environmental water samples by flameless atomic absorption, *At. Absorpt. Newsl.,* 15, 29, 1976.
53. **Baird, R. B., Pouriau, S., and Gabrielian, S. M.,** Determination of trace amounts of seleium in wastewaters by carbon rod atomization, *Anal. Chem.,* 44, 1887, 1972.
54. **Henn, E. L.,** Determination of selenium in water and industrial effluents by flameless atomic absorption, *Anal. Chem.,* 47, 428, 1975.
55. **Martin, T. D., Kopp, J. F., and Ediger, R. D.,** Determining selenium in water, wastewater, sediment and sludge by flameless atomic absorption spectroscopy, *At. Absorpt. Newsl.,* 14, 109, 1975.
56. **Metals Section, Central Regional Laboratory, Region V, U.S. Environmental Protection Agency,** The Determination of Antimony, Arsenic, Beryllium, Cadmium, Lead, Selenium, Silver, and Tellurium in Environmental Water Samples by Flameless Atomic Absorption, Report PB-269 902, National Technical Information Service, Springfield, Ill., 1977.
57. **Van Loon, J. C., Lichwa, J., Rattan, D., and Kinrade, V.,** The determination of heavy metals in domestic sewage treatment plant wastes, *Water Air Soil Pollut.,* 2, 473, 1973.
58. **Carrondo, M. J. T., Perry, R., and Lester, J. N.,** Influence of conditioning agents on the determination of metallic content of sewage sludge by atomic-absorption spectrophotometry with electrothermal atomisation, *Analyst (London),* 104, 937, 1979.
59. **Briese, L. A. and Giesy, J. P.,** Determination of lead and cadmium associated with naturally occurring organics extracted from surface waters using flameless atomic absorption, *At. Absorpt. Newsl.,* 14, 133, 1975.
60. **Standing Committee of Analysts,** Department of the Environment, Mercury in Waters, Effluents and Sludges by Flameless Atomic Absorption Spectrophotometry, Methods for the Examination of Waters and Associated Materials, Her Majesty's Stationery Office, London, 1978.
61. **Allen, H. E. and Minear, R. A.,** Metallic ions, in *Examination of Water for Pollution Control, Vol. 2, Physical, Chemical and Radiological Examination,* Suess, M. J., Ed., Pergamon Press, Oxford, 1982, 43.
62. **Farey, B. J., Nelson, L. A. and Rolfe, M. G.,** A rapid technique for the breakdown of organic mercury compounds in natural waters and effluents, *Analyst (London),* 103, 656, 1978.
63. **Webster, J.,** The estimation of arsenic concentrations in sewage sludges from the Lothian region, *Water Pollut. Control,* 79, 405, 1985.
64. **Sinemus, H. W., Melcher, M., and Welz, B.,** Influence of valence state on the determination of antimony, arsenic, bismuth, selenium and tellurium in lake water using the hydride AA technique, *At. Spectrosc.,* 2, 81, 1981.

65. **Farey, B. J. and Nelson, L. A.**, Water and effluents in *Atomic Absorption Spectrometry, Techniques and Instrumentation in Analytical Chemistry,* Vol. 5, Cantle, J. E., Ed., Elsevier, Amsterdam, 1982, 67.
66. **Webber, M. D. and Gaynor, J. D.**, Extractable metals in mixtures of soil and sewage sludge, in *Proc. Conf. Sludge Handling and Disposal,* Ministry of the Environment, Ontario, 1974, 249.
67. Department of the Environment, Analysis of Raw, Potable and Waste Waters, Her Majesty's Stationery Office London, 1972.
68. American Public Health Association, American Water Works Association, Water Pollution Control Federation, *Standard Methods for the Examination of Water and Wastewater,* 14th ed., American Public Health Association, Washington, D.C., 1976.
69. Princeton Applied Research Corporation, Differential Pulse Stripping Voltammetry of Water and Waste Water, Application Brief W-1, Princeton Applied Research Corporation, Princeton, 1976.
70. **Slavin, W.**, Atomic absorption spectroscopy the present and future, *Anal. Chem.,* 54, 685A, 1982.
71. **Smith, R.**, Organisation and evaluation of interlaboratory comparison studies among Southern African water analysis laboratories, *Talanta,* 31, 537, 1984.
72. **Broekaert, J. A. C. and Leis, F.**, An injection method for the sequential determination of boron and several metals in wastewater samples by inductively-coupled plasma atomic emission spectrometry, *Anal. Chim. Acta,* 109, 73, 1979.
73. **Schramel, P., Li-Qiang, Z., Wolf, A., and Hasse, S.**, ICP-Emissionsspektroskopie: ein analytisches Verfahren zur Klarschlamm-und Bodenuberwachung in der Routine, *Fresenius Z. Anal. Chem.,* 313, 213, 1982.
74. **Oake, R. J., Booker, C. S., and Davis, R. D.**, Fractionation of heavy metals in sewage sludges, *Water Sci. Technol.,* 17, 587, 1984.
75. **Steinnes, E.**, The place of neutron activation analysis in environmental research, *Fresenius Z. Anal. Chem.,* 317, 220, 1984.
76. **Segebade, C., Schmitt, B.-F., Fusban, H. U., and Kuhl, M.**, Application of photon activation analysis to the determination of the distribution of toxic elements in soil of a sewage farm, *Fresenius Z. Anal. Chem.,* 317, 413, 1984.

Chapter 5

CHEMICAL SPECIATION OF HEAVY METALS IN SEWAGE SLUDGE AND RELATED MATRICES

D. L. Lake

TABLE OF CONTENTS

I.	Introduction	126
II.	Discrete Chemical Extractions	127
	A. Discrete Aqueous Extractions	128
	B. Discrete Chloride Extractions	129
	C. Discrete Extractions by Acid Reagents and Chelating Agents	129
	D. Progressive Acidification Techniques	132
III.	Sequential Chemical Extractions	136
	A. Analytical Considerations	136
	1. Technique Descriptions and Choice of Reagents	136
	2. Problems Associated With the Use of Sequential Extraction Techniques	140
	B. Heavy Metal Distributions in Sewage Sludges and Related Matrices	143
References		148

I. INTRODUCTION

Analytical methods for measuring the total metal content of sewage sludges and related matrices are well established.[1-5] However, while total concentrations of metals indicate the extent of contamination, they give little insight into the forms in which the metals are present in sludge or their potential for mobility and bioavailability upon dispersal in the environment. This can only be attained from a detailed knowledge of the speciation of metals in the sludge itself and the changes in speciation likely to occur following disposal.[6-8]

According to Gould and Genetelli,[9] metals may occur in sewage sludges as soluble, adsorbed, organically complexed, precipitated, coprecipitated in metal oxide, or residual forms. Such terms are very ill defined, with the definition of one often overlapping that of another. However, soluble metals may exist in solution as simple ionic forms, or complexed to soluble organics and inorganics.[9] A metal complex is formed by association between an electron-deficient metal atom or ion and an electron-rich species called a ligand, which is either an anion (e.g., Cl^-, OH^-, HCO_3^-, SO_4^-) or a polar molecule (e.g., NH_3).[10] Complexation is principally by coordinate covalent bonding; this is established through donation of a pair of electrons from the ligand to the metal-ion acceptor. However, ionic bonding by electrostatic attraction also plays a major role in many metal-ligand complexes.[10-12] If two or more electron-donating atoms are present within a ligand, they may coordinate to the same metal ion to form a heterocyclic ring termed a *chelate*.[10] Complexes with multidentate ligands (more than one donor atom) are generally more stable than those with monodentate ligands.[12] Humic acids and fulvic acids are examples of multidentate ligands found in sludges.[12] The ability of these compounds to form stable complexes with metal ions is undoubtably due to their high content of the electron-donating ligands RCOOH, phenolic-, alcoholic-, and enolic-OH, and to a lesser extent, RCO and RNH_2.[11] Schnitzer[13] reported 60% by weight of a fulvic acid to be composed principally of carboxyl functional groups. According to Stevenson and Ardakani,[11] metal-fulvic acid complexes are of more significance in the soluble phase than metal-humic acid complexes, with the high acidities and relatively low molecular weights of the former complexes rendering them highly soluble. In addition, individual biochemical molecules (e.g., aliphatic acids and amino acids) are important in the formation of soluble organometallic complexes.[11]

Adsorption may generally be defined as the adhesion, in an extremely thin layer, of gas molecules, dissolved substances, or liquids to the surface of solids with which they are in contact.[14] Adsorption may occur through physical, chemical, or ion-exchange processes.[15] Physical adsorption on the external surface of a particle is based on van der Waals forces or the relatively weak ion-dipole or dipole-dipole interactions (approximately 1 kcal/mol). Additional reactions could occur with physical adsorption on the inner surfaces of particles.[15] Chemical adsorption is characterized by the formation of chemical associations or bonds between ions or molecules from solution and the surface of particles, principally of a covalent nature (i.e., bonds formed by sharing of electrons by two atoms).[14] Adsorption based on ion exchange is a chemical process in which a negative or positive charge on a particle surface is equalized by ions possessing opposite charges.[15] Most particle surfaces have a negative electrical charge; thus in solution, an equivalent number of cations will gather around the particle, where by an electric double layer occurs. Cations entering into exchange reactions are thus held near the charged surfaces by electrostatic attraction or coulombic forces.[14] By definition,[16] cations held on negatively charged sites on the surface of colloids or particles are easily exchangeable with the soil solution and are frequently referred to as "exchangeable" metals. Clay minerals and organic matter, in addition to hydrous oxide materials, all contribute adsorption sites.[14] Adsorption is affected by pH, Eh, ionic strength, and the composition of the solution phase in addition to the clay, oxide, and organic-matter content of the sludge.[17,18]

Organically complexed metals incorporate those forms which are bound to insoluble organic matter, in addition to components of living cells, their exudates, and a spectrum of degradation products, by simple complexation or chelation.[19] Research relating to the mechanisms involved in complexation and chelation between metal ions and solid-phase organic compounds of sewage sludges is limited, but the principles involved are likely to be similar to those described previously for solution-phase organometallic interactions. The metals in soil that occur in insoluble combinations with organic matter are largely those that are bound to components of the humic fractions, particularly humic acids.[11] Since a major proportion of the sludge organic fraction consists of humic compounds, these may also play an important role in the formation of insoluble organometallic complexes in this matrix.

Precipitated metal forms are defined as insoluble substances formed in solution as the result of a chemical reaction,[20] and include metal hydroxides, carbonates, phosphates, and sulfides. Metals may also be coprecipitated, under aerobic conditions, with manganese and iron oxides which may occur as concretions, stains, or coatings on the surfaces of sludge particles. Finally, residual metals may occur as ions inertly bound in crystal lattices of highly stable primary and secondary minerals.[21]

The distribution of metals between the specific forms varies widely according to the chemical properties of the individual metal and the characteristics of the sludge, which are a function of the physical and chemical properties imposed by the particular sludge-treatment processes. These include such parameters as pH, temperature, oxidation-reduction potential, the presence of complexing agents, and the concentrations of precipitant ligands.[22]

Segregation of heavy metals in sewage sludges into all-specific physicochemical forms is not possible, however, with current analytical techniques. This is due not only to the limitation of analytical techniques available as regards constraints imposed by interferences, selectivity, and sensitivity, but also to the complex nature of sewage sludge samples. Valuable information on the partitioning of heavy metals in such a complex matrix into several component fractions has been obtained, however, using chemical-extraction techniques based on selective chemical reagents. Such techniques may either be of a discrete nature, employing a single selective reagent to release a specific metal fraction from within the sludge sample or, alternatively, may incorporate several reagents of increasing extraction strength to release a number of different metal forms in sequence. This fundamental classification into discrete and sequential types provides a basis for discussion of chemical-extraction techniques.

II. DISCRETE CHEMICAL EXTRACTIONS

Discrete chemical extractions involve either simple shaking of a sludge sample in a reagent at a known sludge solids-to-volume ratio over an optimum time period[23-30] or continuous leaching of columns of sludge.[31,32] Many of the discrete extraction schemes applied to sewage sludge matrices were developed by modifying techniques originally applied to soils, in which metal cations similarly exist in several different physicochemical forms. Such methods are simple and rapid, but the chemical reagents used may themselves alter the indigenous speciation of the sludge samples by causing an immediate perturbation of equilibrium conditions.[8] In addition, such techniques suffer from the difficulty of finding a single reagent effective in dissolving quantitatively the desired metal form without attacking others.[33] Isolated metal forms may not necessarily be representative of definable physical or chemical forms but can only be defined operationally by the method of extraction.[8]

The different concentrations and types of reagents used to extract a particular metal form renders the choice of an optimum extractant from a compilation of published data difficult, while uncertainty as to the significance of the extracted metal limits the usefulness of such comparisons. This has been acknowledged by Latterell et al.[34] who, although finding sodium pyrophosphate ($Na_4P_2O_7$) to be the most efficient extractant of plant-available metal forms, preferred to use the more commonly employed reagent diethylenetriaminepentaacetic acid (DTPA) because the results thus obtained could then be directly compared to others reported in the literature. Furthermore, the consistency of extraction with the individual reagents may vary, as demonstrated by an interlaboratory comparison in which the accuracy of methods involving either ethylenediaminetetraacetic acid (EDTA) or acetic acid as the extractant differed considerably.[35]

Despite the shortcomings of the method, however, numerous investigations into the physicochemical nature of heavy metals in sewage sludges and related matrices have been undertaken utilizing chemical single-extractant techniques. Such studies are often restricted to the metals cadmium, copper, nickel, lead, and zinc since these are the contaminants of principal concern when sludge is applied to land.[36,37]

A. Discrete Aqueous Extractions

Simple aqueous extractions have frequently been employed to investigate the occurrence of water-soluble heavy metals in sewage sludges. By definition, water-soluble forms comprise metals which exist in ionic, molecular, or colloidal form in the sludge-solution phase[38] and hence are most available for plant uptake.

Extraction of air-dried digested sewage sludge with tap water resulted in <1% of lead and approximately 2% of copper being leached, while up to 11% of cadmium and 36% of zinc were water soluble.[32] Jenkins and Cooper[31] found similar aqueous extractability for copper (0.3%) following repeated percolation of columns of dried digested sludge with distilled water, but that for zinc (1.7%) was much less. Nickel extractability was greater at 14.3%. Reductions in the water solubility of these metals following successive percolations were attributed to the development of anaerobic conditions.[24] Similar consistently low water solubilities following anaerobic incubation for 1 week have been reported for cadmium (<0.1%) and lead (<0.03%), while copper, nickel, and zinc solubilities were <0.1 to 6.2%, 1.5 to 9.2%, and <0.01 to 0.8%, respectively.[26] Reductions in the water solubility of copper from 0.6 to 0.14% and nickel from 4.9 to 2.8% following anaerobic incubation for 3 months have been reported by Bloomfield and Pruden[24] although cadmium, chromium, and lead increased in solubility from <0.1 to 7%, <0.01 to 0.8%, and <0.2 to 0.6%, respectively. The solubility of zinc remained below 1%. Subsequent aeration increased the water solubility of chromium, copper, and zinc, but had no corresponding effect on cadmium, nickel, and lead.[24]

It was apparent from these observations that, in general terms, cadmium and nickel were the most (and lead the least) water-soluble metals in sludge. Considerable variation in metal solubility between sludges has been reported, however, reflecting differences in the properties of the sludge samples employed. For example, the considerable difference in the percentages of water-soluble zinc extracted from the two digested sewage sludges investigated by Lagerwerff et al.[32] (namely <0.2% as compared to 36%) was probably due to the much higher indigenous pH value of the former sludge (pH 6.5 compared to pH 4.0, respectively). Although water-soluble species have generally been found to constitute <10% of the total metal content of a sludge, they are of particular significance in representing the metal forms most immediately relevant to plant uptake.

With respect to related matrices, Bradford et al.[25] observed that concentrations of cadmium, cobalt, copper, nickel, and zinc in saturation extracts of air-dried sewage sludge were consistently higher than those obtained from soils. In addition, water-soluble forms of copper, nickel, lead, and zinc removed in aqueous extracts of sludge-soil mixtures have been found to be negligible by Jenkins and Cooper[31] and Petruzzelli et al.[39] According to Bloomfield and Pruden,[24] the addition of soil to sludge (5:1) reduced the amounts of copper, nickel, lead, and zinc removed by aqueous extraction by 80, 60, 20, and 70%, respectively, compared with sludge alone. In contrast, however, the water solubility of cadmium and chromium was increased by approximately 50%.[24]

B. Discrete Chloride Extractions

Calcium, barium, and magnesium chloride solutions, which have frequently been used to extract exchangeable forms of metals from sediments[40,41] and soils,[21,42-44] have also been applied to sewage sludges.[29,30,32] The mechanism of extraction involves displacement of metals loosely held on negatively charged exchange sites by calcium, barium, or magnesium ions, which saturate these sites. A comparison of these chloride reagents with nitrates and neutral acetate salts, which are commonly used to extract exchangeable metal forms from soils,[45-47] suggested efficient extraction of metals from a sludge-derived soil.[48] In support of this, McLaren and Crawford[21] considered calcium chloride ($CaCl_2$) to be preferable to the frequently used neutral N ammonium acetate (NH_4OAc) in extracting exchangeable cations from soils since it has less effect on the natural soil pH and extracts less organic matter. Objections have been raised, however, to the use of chloride extractants for determining exchangeables due to the formation of dissolved metal-chloro complexes.[16]

Lagerwerff et al.[32] found that while <2% and <1% of total copper and lead, respectively, were exchanged from columned-digested sludge of pH 4.0 by consecutive leachings with 0.06 M $CaCl_2$, extractability of cadmium and zinc was greater at 46 and 81% respectively. Comparable behavior of cadmium and zinc relative to that of copper and lead was more recently found following four repeated extractions of air-dried raw, activated, and digested sludges with 0.05 M $CaCl_2$ (pH 5.3), although concentrations removed were generally 0.1 to 10% of the total metal concentrations determined.[30] Similar observations were made by Förstner et al.[49] with respect to exchangeable metal forms in activated sewage sludges, these authors using the term ''metal pairs'' to denote similarity in speciation behavior of copper-lead and the opposing ion pair zinc-cadmium, respectively.

In the study described earlier, Lagerwerff et al.[32] also investigated a second digested sewage sludge, which had a pH of 6.5. Concentrations of cadmium and zinc displaced from this sludge by $CaCl_2$ represented only 4.6% and 0.4% of respective total concentrations; this is in sharp contrast to the 46% total cadmium and 81% total zinc extracted from the digested sludge of pH 4.0. Such reductions in exchangeable metal were undoubtedly a facet of the higher indigenous pH of the second sludge, and thus the relative competition existing between the metal and hydrogen ions for the exchange sites. Cavallaro and McBride[44] similarly found zinc removal by 0.05 M $CaCl_2$ washes of soil clays to be highly dependent upon pH, with extractability increasing rapidly below pH 5.0. Hence, zinc displaced from artificially metal-loaded sewage sludges by barium chloride ($BaCl_2$) buffered at pH 7.0 was found by Adams[29] to represent only 3.9% of the total, while 4.2% copper and 2.8% nickel comprised easily exchangeable forms. This author presented evidence of a further 48, 67, and 91% of zinc, copper, and nickel, respectively, being associated with exchange sites, but in a complexed form and thus too strongly bound to be displaced by the neutral $BaCl_2$ solution employed. With respect to soils, metals existing as firmly bound ions in exchange complexes are usually extracted by protons from dilute acetic acid solutions.[38]

An assessment of the effect of sludge type on metal extractability in 0.5 M $CaCl_2$ revealed that concentrations of cadmium, copper, nickel, and lead extracted from liquid samples were lower in digested sludge than in raw or activated forms originating from the same treatment works.[30] Lead extractability was negligible in the digested sludge; such stabilization in sludge disposed to land may account for its observed nonreactivity with regard to uptake by plants.[50] In contrast to the other metals studied, zinc extractability remained comparatively constant in all three sludge types.[30] The apparent persistence of zinc in a $CaCl_2$-extractable form concurs with suggestions made by Adams and Sanders[51] that cation-exchange processes are important in zinc binding.

C. Discrete Extractions by Acid Reagents and Chelating Agents

Extraction techniques based on acidification or chelating organics, which have been used

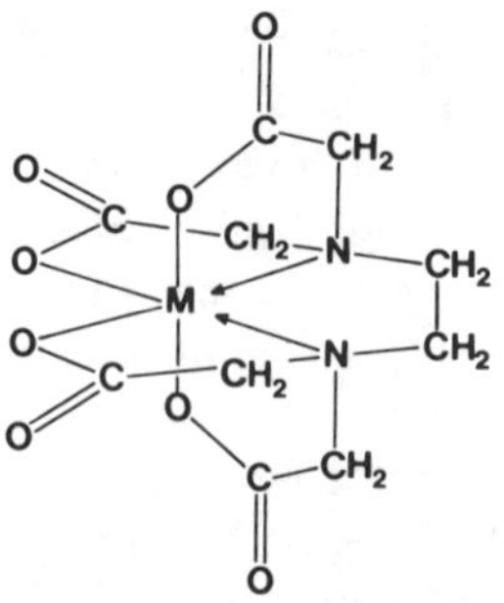

FIGURE 1. Chelation of a metal ion (M) by EDTA.

to determine "available" forms of metals in soils,[52-58] have also been applied to sludges and sludge-soil matrices. The term "available", which has been utilized in micronutrient soil tests to denote the nutrient status of soils with respect to plants, is difficult to define, but may relate to several different metal species, with water-soluble, exchangeable, and organically bound forms being of most significance.[46] These three forms are believed to be in equilibrium with one another, with the equilibrium being affected by pH, Eh, and the concentration of metals and ligands.[59]

Acetic acid, which has also been used to extract metal bound on specific sorption sites from soils,[21] is widely employed as an extractant of an overall "available" component from sludge matrices,[23,24,27,28,60] while other relatively weak acids, such as citric acid, have received only limited application.[31] Ellis and Alloway[61] found the acetic-acid-extractability of cadmium, lead, and nickel in particular to be useful in determining concentrations of these metals likely to be taken up by plants grown on sludge-amended soils. However, in a comparison of reagents used to evaluate plant-available nickel, there was no significant correlation found for citric-acid-extractable metal.[56] Stronger mineral acids have been used at concentrations ranging from 0.01 to 4.0 M H^+ to extract "available" metals from sludges.[25,26,32] These tend to be efficient extractants, but doubt has been expressed with regard to the suitability of such reagents for determining plant-available metals, since the pH achieved during extraction is rarely representative of that of the sludge or sludge-amended soil.[27] In support of this, Sauerbeck and Rietz[48] suggested that weakly acidic but well buffered extracting agents were to be preferred as indicators of availability since strong unbuffered reagents gave unrealistically large recoveries and were unpredictable in their extent of neutralization during the extraction procedure.

EDTA and DTPA are the most frequently used chelating organics in the determination of "available" forms of metals in sludges[24,62] and sludge-amended soils.[63-67] Extraction based on chelation involves the coordination of two or more donor atoms present in the chelating ligand to the same metal ion in such a way so as to form a heterocyclic ring termed a chelate.[12] EDTA, for example, is a sexadentate ligand (ie., contains six donor atoms capable of coordination to a single metal ion) which forms five chelate rings when all donor atoms are coordinated to a metal, as illustrated in Figure 1.

Chelating organics have been shown to be efficient extractants of "available" metal forms, with rankings such as 0.05 M EDTA >0.5 M acetic acid >1.0 M NH_4OAc being cited.[28] Such studies tend, however, to base extraction efficiency only on the total concentration of metal released by reagents commonly used to measure plant availability from soils, completely neglecting crucial studies on plant uptake. Good correlations between concentrations of metals extracted by acetic acid, EDTA, and DTPA and concentrations taken up by plants from sewage sludge-amended soils have been reported,[61,63,65,68,69] but the degree and form of correlation varies depending on the metal and soil, climate, and plant factors, as further discussed in Volume II, Chapter 5. In addition, the validity of equating extractability by such reagents with availability has been questioned since such extractants tend to dissolve forms which are not available and/or do not dissolve forms which are available.[69] It is known, for example, that EDTA has the ability to extract oxide-occluded[70] and sulfide-precipitated[26] forms of some metals, which are not readily available to plants. When framing guidelines it may thus be unwise to equate "extractable" to "available" metal forms.

Despite the obvious drawbacks of the method, extraction of potentially toxic metals present in sludge, sludge-soil mixtures, and soils by acid reagents and chelating organics continues to be widely used as an approximation of their uptake by plants grown on these matrices. A wealth of information regarding the comparative solubility of various metals present in sewage sludge and sludge-soil mixtures in such reagents has thus been established. Between 50 and 75% of copper, nickel, and zinc have been extracted by percolating $0.095\ M$ citric acid, buffered at pH 2 to 6, through columns of digested sludge,[31] while nickel and zinc, in addition to manganese, have also been found to be highly soluble in $0.42\ M$ acetic acid, with over 95% of total zinc being present in readily soluble forms in some cases.[23] Acetic-acid-extractable forms of chromium, copper, and lead, however, represented on average only 3.1, 6.9, and 2.8% of the total respective metal.[23] Similarly, extractabilities of copper and lead in $0.5\ M$ hydrochloric acid were low compared to those of cadmium, nickel, and zinc, which represented 69, 59, and 73% of respective total concentrations.[26] Such findings suggest similarity in the mechanisms by which copper and lead are retained in sludge while different mechanisms are likely to be responsible for the retention of more easily extractable metals such as cadmium, nickel, and zinc.

The effect of anaerobic incubation of digested sewage sludge on metal extractability in $0.5\ M$ acetic acid and $0.05\ M$ EDTA has been investigated by Bloomfield and Pruden.[24] Under such conditions, the proportion of the total cadmium, copper, nickel, and zinc extracted by acetic acid decreased from 60 to 50%, 10 to 0.2%, 39 to 5%, and 60 to 20%, respectively, while acetic-acid-extractable chromium and lead increased from 0.9 to 4.6% and 1.2 to 5.7%, respectively. Similar behavior was observed for EDTA-extractable cadmium, chromium, copper, lead, and zinc, although concentrations released were always higher than those extracted by acetic acid. With respect to nickel, anaerobic incubation had the effect of increasing its extractability in EDTA. Subsequent aeration generally increased the acetic acid and EDTA extractability of metals. However, under such conditions, acetic-acid-extractable chromium was reduced to 3% while EDTA-extractable chromium was hardly affected.[24] This change in extractability in specific metals under different redox conditions can be explained by the changes in metal form which occur. For example, the increase in the extractability of lead by EDTA under anaerobic conditions could result from an increase in lead carbonate which is readily soluble in EDTA.[26]

Discrete extraction techniques based on acidification and chelating organics have yielded comparative solubility data with respect to soils and sludge-amended soils. Sewage sludges have generally been found to contain a greater proportion of their total metal content in forms extractable by $0.42\ M$ acetic acid in comparison with soils.[23] For example, median values of "available" copper and zinc in sludge were 140 and 500 times those in soils, respectively. Concentrations of $0.5\ M$ acetic-acid-extractable chromium and copper and $0.05\ M$ EDTA-extractable nickel and zinc have also been reported as being higher in sludges than in sludge-amended soils,[24] while extractabilities of copper by EDTA and of nickel and zinc by acetic acid in sludge-treated soils were found to be greater than extractabilities of the native metals in soils.[67] Similarly, cadmium, copper, nickel, lead, and zinc in soils generally show enhanced DTPA-extractability following sludge application.[63-66] Other metals such as cobalt and chromium appear to be largely unaffected, however.[63] Using DTPA as an extractant of "available" metal, Williams et al.[71] reported a high percentage of cadmium, copper, lead, and zinc added to soils in sewage sludge to be available, while cobalt, manganese, and nickel were found to be moderately to slightly available and iron and chromium relatively unavailable. Iron and chromium as trivalent ions tend to precipitate, following hydrolysis, to insoluble hydroxides and oxides.[72] Although the general consensus is that the addition of sludge to soils has the immediate effect of increasing the "available" metal pool, it has been suggested that over longer time periods there may be a tendency for metal forms to stabilize.[62,64] Further applications of such extraction techniques in elucidation of the changes that occur in metal form following sludge disposal to agricultural land are discussed in Volume II, Chapter 5.

D. Progressive Acidification Techniques

In addition to providing an insight into the characterization of metals in sludges and sludge-soil mixtures, chemical-extraction techniques may further serve to elucidate the behavior of metals following changes in environmental conditions. One of the most influential parameters controlling the transference of metals from immobile solid-phase forms to more mobile, and possibly more bioavailable, solution-phase forms is pH.[51,73-76] Adsorption equilibria, the stability of organomineral complexes, and the redox potential are particularly affected by this parameter.[47,77]

Interlaboratory investigations into the effects of pH on metal solubility in sludge-polluted soils have been reported by Cottenie[73] and Kiekens and Cottenie,[78] who observed the mobility of cadmium, cobalt, chromium, copper, nickel, lead, and zinc resulting from progressive acidification of soil-water suspensions. Experimentally, the technique of progressive acidification incorporated 10-g samples of air-dried sludged soil being suspended in 30 mℓ distilled water and the pH adjusted to the original value of the soil. Fresh soil-water suspensions were simultaneously prepared at pH values of 4.0, 2.0, or 0.5. The samples were then stirred continuously for 30 min during which time adjustments to maintain the pH at ambient, 4.0, 2.0, or 0.5 were made by controlled addition of nitric acid at concentrations ranging from 0.1 to 4.0 M, depending on the buffering capacity of the sludge-soil mixture. After equilibration, the final volume of the suspensions was made up to 50 mℓ by addition of distilled water. This was followed by centrifugation of the suspensions at 3000 g for 15 min and filtration of the resultant supernatants. Filtrates were subsequently measured for pH prior to determination of heavy-metal concentrations by atomic absorption spectrophotometry.[73,78] The concentrations of individual metals in the filtrate at a given pH corresponded with the mobile fraction under these conditions. Such a technique thus distinguishes between a metal fraction which is soluble at actual soil conditions (corresponding with the so-called "nutrient intensity" in the case of essential elements) and a further mobilizable fraction which could pass into solution under conditions of changing soil pH as a consequence of desorption of cations, dissolution of solids, and decomposition of organomineral complexes.[79]

Consistent mobilization patterns were obtained for each metal studied by Cottenie[73] and Kiekens and Cottenie[78] as a result of progressive acidification of sludge-polluted soils, depending essentially on the nature of the individual elements. Thus, a definite break-point or threshold pH was identified for each metal, above which the percentage of mobilization was relatively low and stable, and below which mobilization tended to increase exponentially. Similar patterns of mobilization were observed for lead and copper, which tended to pass into solution at approximately pH 2.0, while cadmium, cobalt, nickel, and zinc exhibited greater mobility with a threshold of mobilization nearer to pH 4.0 (Figure 2). With respect to chromium, however, no significant mobilization was apparent as a result of decreasing pH.[78]

Modified versions of the acidification technique developed by Cottenie[73] have been employed by others to investigate the solubilization of metals in sewage sludge[29,30,51,80,81] and sludge-soil[48,80,82] matrices under conditions of decreasing pH. The threshold pH values obtained for mobilization of a number of metals during these investigations are compared

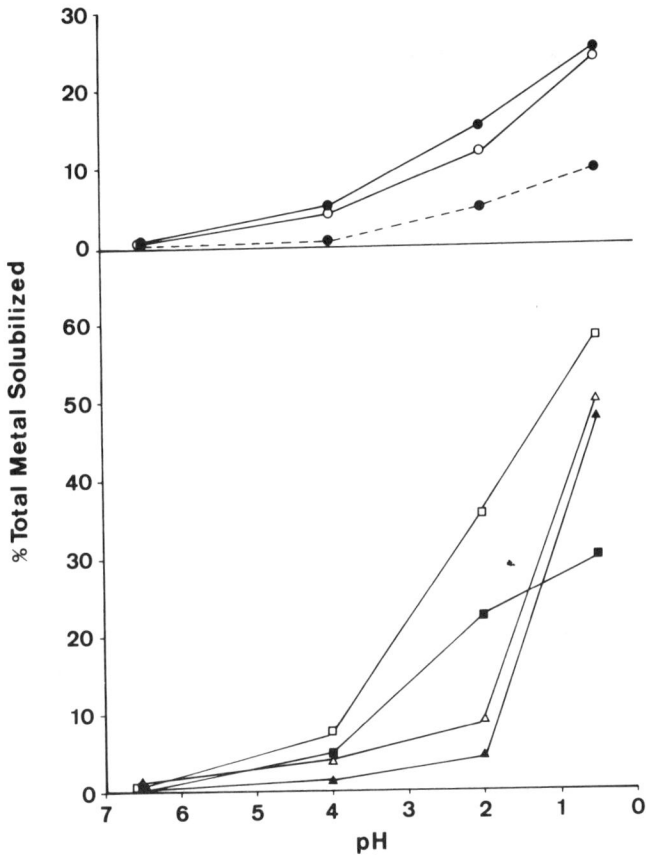

FIGURE 2. Solubilization of cadmium (□), cobalt (-●-), chromium (--●--), copper (△), nickel (○), lead (▲) and zinc (■) from a sludge-polluted soil by progressive acidification. (Adapted from Kiekens, L. and Cottenie, A., in *Proc. 3rd Int. Symp. Processing and Use of Sewage Sludge*, L'Hermite, P. and Ott, H., Eds., D., Reidel, Dordrecht, the Netherlands, 1984. With permission.)

in Tables 1 and 2. Fairly good agreement is apparent between the separate studies, with copper and lead emerging as the metals most resistant to and nickel and zinc as the metals least resistant to the effects of decreasing pH in both sludge and sludge-soil matrices. The behavior exhibited by the individual metals is linked to their physicochemical characteristics and the different mechanisms governing their retention in these matrices. According to Adams and Sanders[51] and Kiekens and Cottenie,[78] the increased concentrations of nickel and zinc occurring in solution as a consequence of decreasing pH are controlled by adsorption-desorption equilibria (including cation-exchange processes). McBride and Blasiak[83] concluded that specific adsorption on hydrous oxides of aluminum and iron is the major process determining zinc solubility in soil below pH 6.5, but in sludges, organic adsorption is likely to be more important. Copper and lead, however, are known to exhibit less reactivity to both decreasing pH and increasing ionic strength in comparison to metals such as cadmium and zinc.[84] It has been suggested that the former metals are largely associated with organic materials over a wide pH range in soils,[21,85-87] which must be even more so in sludge which is principally organic matter. Both copper and lead form stronger complexes with organic matter than either nickel or zinc.[11,85,86] According to Adams and Sanders,[51] this results in the threshold value below which copper and lead are released to be lower than that for nickel

Table 1
COMPARATIVE THRESHOLD pH VALUES OBTAINED FOR MOBILIZATION OF METALS FROM SEWAGE SLUDGES BY PROGRESSIVE ACIDIFICATION

Metal	Metal-enriched filter-pressed raw sludge[80]	Thawed, metal-enriched, filter-pressed raw sludge[51]	Thawed, filter-pressed raw sludge[29]	Raw, activiated, and digested sludge[30,81]	
				Liquid	Dried
Cd	—	—	—	2.0	4.0
Cu	4.0	4.5	<4.5	No mobilization	2.0—4.0
Ni	6.0	6.3	6.4	No threshold	No threshold
Pb	—	—	—	2.0	4.0
Zn	6.0	5.8	6.1	4.0	6.0

Table 2
COMPARATIVE THRESHOLD pH VALUES OBTAINED FOR MOBILIZATION OF METALS FROM SLUDGE-AMENDED SOILS BY PROGRESSIVE ACIDIFICATION

Metal	Reference 78	Reference 79	Reference 48	Reference 80	Reference 82
Cd	4.0	5.0	4.0	—	—
Co	4.0	—	—	—	—
Cr	No mobilization	—	—	—	—
Cu	2.0—4.0[a]	3.0	—	4.0	4.5—5.0[a]
Ni	4.0	5.0	—	6.0	6.2—6.5[a]
Pb	2.0	—	2.0	—	—
Zn	4.0	5.0	5.0	6.0	5.8—6.5[a]

[a] Threshold pH values varied according to soil type.

and zinc since about 80% of the sludge organic matter remains insoluble throughout acidification. An increase in the mobilization observed for copper and lead is thus likely to result primarily from the direct solubilization of the organically bound fraction.[81]

Although the concentrations of metal released from sludge-treated soils have been found to be less than those from sludge alone,[82] the similarity of threshold values observed for specific metals (Tables 1 and 2) suggests that mechanisms governing their physicochemical behavior are common to both types of matrix. In addition, variations according to sludge type are not apparent between raw, activated, and digested sludges.[30,81] However, variations in the mobilization patterns of individual metals according to sludge form[30,81] (i.e., whether in a liquid or air-dried state) and soil type[79,80,82] have been reported. It has been observed that air-dried samples of sewage sludge tend to release metal at higher pH values than corresponding liquid samples when subjected to acidification, as illustrated in Figure 3.[30,81] It may be that this phenomenon was caused by the differing particulate sizes of the liquid and dried sludges and hence the differing surface areas exposed for adsorption-desorption reactions. Alternatively, such an effect may have resulted from the storage of liquid sludge since a transfer of metal species from easily extractable (and hence relatively mobile) forms to forms more resistant to the effects of decreasing pH have been observed under such conditions.[88] With respect to discrepancies according to soil type, Adams and Sanders[82] found slightly higher threshold values of mobilization for copper, nickel, and zinc from two sandy loams than from a clay loam as indicated in Table 3. Cottenie et al.[79] reported similar discrepancies between clay and sandy loam soils. Such variation may be partially explained

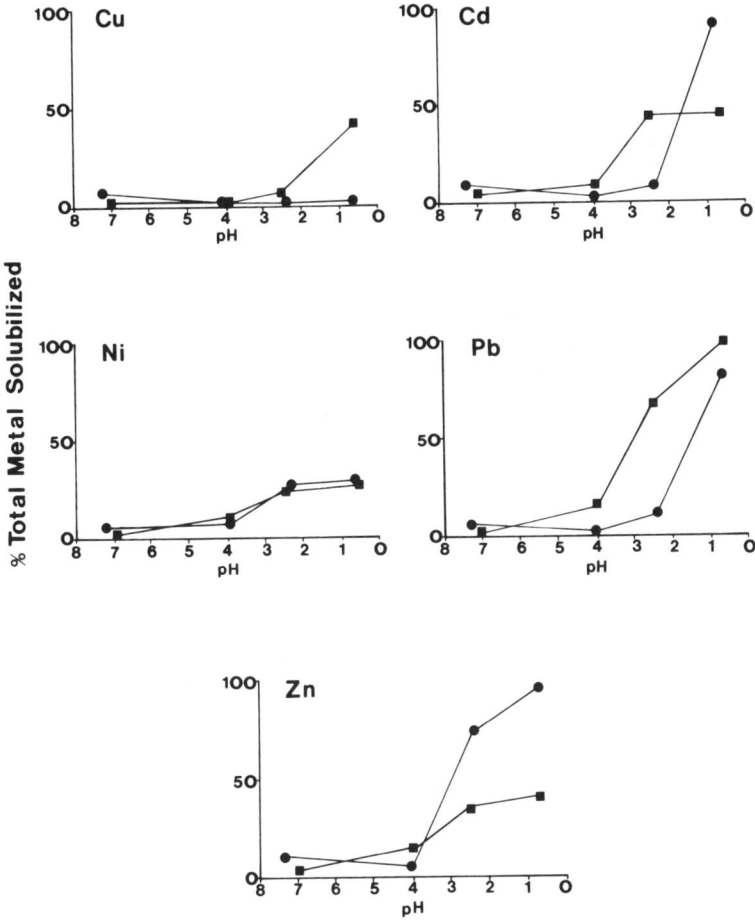

FIGURE 3. Percentage metal solubilized by progressive acidification of air-dried (■) and liquid (●) forms of an activated sludge sample. (From Rudd et al.[30] and Lake[81]).

Table 3
APPARENT THRESHOLD pH VALUES FOR SOILS TREATED WITH METAL-LOADED SLUDGES

	Soil type		
Metal	Sandy loam 1 (Cottenham series)	Sandy loam 2 (Newport series)	Clay loam (Carswell series)
Cu	5.0	5.0	4.5
Ni	6.5	6.5	6.2
Zn	6.5	6.3	6.0

Adapted from Adams, T. McM. and Sanders, J. R., in *Proc. Int. Conf. Environmental Contamination*, CEP Consultants Ltd., Edinburgh, 1984. With permission.

by the difference in texture between the soils, since the heavier clay loams retain metals more strongly than light-textured sandy loams. Variation between the concentraions of copper, nickel, and zinc released by the two sandy loam soils investigated by Adams and Sanders[82] at a given pH below the respective threshold values suggested, however, that differences in cation-exchange capacity (to which organic matter will contribute) and the concentration of reactive oxides, both of which contribute to the adsorption surface, are better indices of metal-retaining ability than is texture alone. Soil properties will thus have a large influence on the solubilities of metals derived from sewage sludge.

The threshold pH values of metal mobilization obtained from sewage sludges have practical implications for the spreading of sludge on agricultural land. In England and Wales, it is recommended that soil pH be maintained above 6.0 for grassland and 6.5 for arable soils to which sludge is applied;[89] many countries have similar recommendations.[90] The threshold values reported for zinc and particularly for nickel in sewage sludges approach these recommended values (Table 1), suggesting that these may be limiting metals and the most likely to cause phytotoxic effects following sludge disposal to agricultural land. This is of particular concern since factors not considered in guidelines, such as soil properties and rhizosphere effects (including carbon dioxide respiration by roots and associated microorganisms, in addition to the production of acidic organic exudates), may further reduce the pH in the vicinity of plant roots, resulting in increased uptake of nickel and zinc by plants.[51,82] Such diverse factors create difficulties in making simple guidelines for the disposal of metal-contaminated sludge to land. The threshold values obtained for copper and lead (Table 1) indicate that these metals will be less of a problem in normal agricultural soils. Since, however, copper, nickel, and zinc are usually present together in natural sludges, the highest threshold of the three should be regarded as the minimum soil pH at which metal-contaminated sludge can be applied to agricultural soil.[51]

III. SEQUENTIAL CHEMICAL EXTRACTIONS

A. Analytical Considerations

Sequential chemical-extraction techniques are considered to be of greater value than techniques based on a single extractant in determining metal distribution in wastewater sludge;[26] although more time consuming, they provide additional information on the origin, mode of occurrence, biological and physicochemical availability, mobilization, and transport of heavy metals.[40] Such techniques initially involve extraction of a dried sludge sample by shaking over a predetermined time period with an appropriate volume of a relatively weak reagent, followed by centrifugation. The resultant supernatant is subsequently decanted, and possibly filtered,[6,88,91] in preparation for metal analysis, while the remaining sludge pellet is then resuspended and washed in water prior to extraction by another reagent.[26,91] This procedure is followed for a sequence of reagents of progressively increasing strength. Finally, the "residual" metal is extracted from the sludge pellet using mineral acids, at ambient or increased temperatures, or homogenization techniques. In those procedures where washing is not included between extraction steps, computer calculations have been developed to compensate for metal "carry-over".[6,36,92] Determination of heavy-metal concentrations of the sequential extracts and the final pellet extract is by various analytical techniques. These include atomic absorption spectrophotometry,[6,26,36,91,93] X-ray fluorescence spectrocopy,[88] or DC argon plasma atomic emission spectrophotometry.[94] Standard metal solutions prepared in the respective reagents are used to compensate for the background matrix interference.

1. Technique Descriptions and Choice of Reagents

The major sequential extraction schemes applied to wastewater sludges are presented in

Table 4. Reagents utilized were chosen on the basis of their selectivity and specificity towards particular physicochemical forms of heavy metals, although variations in reagent strength, volume, and extraction time between schemes are apparent. While each sequential extraction scheme differs in detail, there are essentially two approaches: either they are modifications of extraction techniques developed for analysis of soils[21,95,96] or, alternatively, they are modifications of those devised for sediment analysis.[40,41,97]

Stover et al.[26] detailed a sequential extraction procedure modified from a technique developed for use on soils by McLaren and Crawford.[21] This technique incorporated 1.0 M potassium nitrate (KNO_3); 0.5 M potassium fluoride (KF) at pH 6.5; 0.1 M $Na_4P_2O_7$; 0.1 M EDTA (pH 6.5), and 1.0 M nitric acid (HNO_3) to fractionate metals in sludges into forms designated as exchangeable, adsorbed, organically bound, carbonate, and sulfide, respectively. The mechanism of extraction by KNO_3 is based on an ion-exchange reaction whereby metals loosely held on negatively charged exchange sites, located on sludge organic and inorganic components, are displaced by potassium ions, which saturate these sites. Extraction of adsorbed metals by KF is essentially through the formation of soluble metal-fluoride complexes. At the pH and concentration of KF used by Stover et al.,[26] minimal solubilization of organically bound metals would be expected to occur. Removal of organically complexed metal by the alkaline $Na_4P_2O_7$ reagent is principally through its ability to solubilize organic matter.[98-101] Pyrophosphate is however a strong inorganic chelating agent under alkaline conditions, and recent electron spin resonance studies[102] suggest that this extractant may cause some heavy metals to dissociate from the organic complexes forming soluble metal-pyrophosphate complexes. Such complexes have previously been reported by Mortvedt and Osborn,[103] Asher and Bar-Yosef,[104] and Bar-Yosef and Asher.[105] Copper, for example, is probably complexed at equatorial ligand positions to pyrophosphate,[102,106] as illustrated in Figure 4. Utilization of $Na_4P_2O_7$ by Stover et al.[26] in preference to the more commonly recommended EDTA reagent[46] was due to the ability of $Na_4P_2O_7$ to extract a greater amount of colloidal organic matter but a lower concentration of metal carbonates than EDTA. On the basis of results obtained from extractions of pure metal precipitates by various reagents, EDTA was found, however, to be a superior reagent for complete yet selective extraction of metal carbonates.[26] Dissolution of metal carbonates by this reagent is principally through the formation of soluble chelates. Solubilization of sulfide forms of heavy metals by HNO_3 is, however, a pH-related phenomenon.

Minor modifications to the technique devised by Stover et al.[26] have been made by Oake et al.[88] and Lake et al.[91] These include the replacement of 1.0 M HNO_3 with 6.0 M HNO_3 for increased efficiency in the extraction of sulfide forms. More major modifications were made by Emmerich et al.,[6] who replaced the KF and $Na_4P_2O_7$ reagents used by Stover et al.[26] with 55.5 M "ion-exchange water" and 0.5 M sodium hydroxide (NaOH), respectively. The preferential use of "ion-exchange water" by these authors was based on the greater efficiency of this reagent in extracting metals adsorbed by hydrous oxide surfaces.[107] In addition, Emmerich[107] questioned the use of KF to extract adsorbed forms as soluble fluoride complexes, considering such complexes to be less stable than comparable hydroxide complexes that might form at the pH of the extracting solution (pH 6.5). Extraction with NaOH is a standard procedure for removal of organics from soils.[101,108] Substitution of $Na_4P_2O_7$ was based on the ability of NaOH to extract a higher percentage of organic matter, in addition to complexed metal, and a lower percentage of carbonate precipitated metal than $Na_4P_2O_7$.[107]

Sequential extraction techniques derived from schemes originally applied to sediment matrices[49,92-94] are more elaborate than those derived from soils (Table 4). Although varying in manipulative complexity, such techniques generally fractionate metals in sludges into the following sequence of operationally defined phases:

1. "Exchangeable" phase. By extraction with NH_4OAc, $BaCl_2$ or magnesium chloride

Table 4
REAGENTS UTILIZED IN THE SEQUENTIAL EXTRACTION OF HEAVY METALS FROM SEWAGE SLUDGES

Sludge types (metals studied are in parentheses)

Designated chemical form	Digested[34] (Cd, Cu, Pb, Zn)	Digested[26] (Cd, Cu, Ni, Pb, Zn)	Raw, activated, digested[88] (Cd, Cr, Cu, Ni, Pb, Zn)	Raw, digested[91] (Cd, Cu, Ni, Pb, Zn)	Digested[6] (Cd, Cu, Ni, Zn)	Digested[92] (Zn)	Activated[49] (Cd, Cu, Pb, Zn)	Digested ash[94] (Al, Cd, Co, Cr, Cu, Fe, Mn, Ni, Pb, Zn)	Digested[93] (Cd, Cu, Ni, Pb, Zn)
Soluble	H_2O				(Footnote[a])				(Footnote[a])
Exchangeable	KNO_3	KNO_3	KNO_3	KNO_3	KNO_3 Ion-exchange	NH_4OAc	NH_4OAc (Footnote[b])	$MgCl_2$ (Footnote[b])	$BaCl_2$
Adsorbed	KF	KF	KF	KF	H_2O	EDTA	H_2O_2-HNO_3 and NH_4OAc (Footnote[c])	H_2O_2-HNO_3 and NH_4OAc (Footnote[c])	H_2O_2-HNO_3 and NH_4OAc (Footnote[c])
Organically bound	$Na_4P_2O_7$	$Na_4P_2O_7$	$Na_4P_2O_7$	$Na_4P_2O_7$	NaOH	NaOCl NH_4OAc			
"Available"	DTPA								
Carbonate		EDTA	EDTA	EDTA	EDTA			NaOAc-HOAc	NaOAc-HOAc
Sulfide		HNO_3	HNO_3	HNO_3	HNO_3		$NH_2OH\cdot HCl$-HNO_3	H_2O_2-HNO_3 and NH_4OAc	H_2O_2-HNO_3 and NH_4OAc
Oxide bound[d]						Oxalate[e] mixture (dark and UV)	H_2O_2-HNO_3 and NH_4OAc (Footnote[c]) $NH_2OH\cdot HCl$-HNO_3 and Oxalate[e]	(Footnote[c]) $NH_2OH\cdot HCl$-HOAc	(Footnote[c]) $NH_2OH\cdot HCl$-HOAc and NH_4OAc
Residual	HNO_3					Homogenization HNO_3	$C.HNO_3$	H_2O_2-HF-HCl	HCl-HF

[a] Soluble phase separated from solid phase.
[b] Correct order of extraction scheme is exchangeable, carbonate, oxide-bound, organic/sulfide, residual.
[c] Organic and sulfide forms classed as a single "oxidizable" phase.
[d] Coprecipitated with or occluded in iron and manganese oxides; also designated the "reducible" phase.
[e] Consists of oxalic acid $(CO_2H)_2$ and oxalate $(CO_2NH_4)_2$.

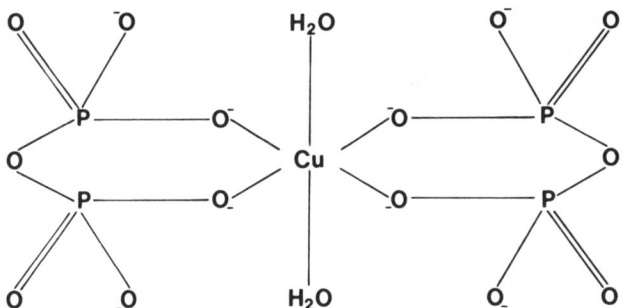

FIGURE 4. Copper-pyrophosphate complex in aqueous solution.

(MgCl$_2$), all at a pH value of 7.0. These reagents contain cations which saturate the exchange sites within the sludge matrix thus causing displacement of heavy metals which are loosely bound to such sites. The formation of heavy-metal complexes with chloride or acetate may, however, affect the efficiency of extraction of "exchangeable" metal forms by these reagents.[16,40,97]

2. "Carbonate" phase. By extraction with sodium acetate (NaOAc) acidified to pH 5.0 using acetic acid (HOAc). Removal of this phase is probably due to both selective dissolution of metal carbonates at pH 5.0 and the formation of soluble complexes between NaOAc and the metals released.[40]

3. "Reducible" phase. Incorporating metals occluded by iron and manganese oxides. Extraction is usually by an acidified reducing agent; hydroxylamine hydrochloride (NH$_2$OH.HCl) coupled to acetic acid (pH 2.0)[109] is most frequently used to extract metals associated with this phase from sewage sludges. Solubilization of iron and manganese oxides by NH$_2$OH.HCl involves reduction to their ferrous and manganous forms, respectively. Acetic acid at pH 2 maintains in solution the heavy metals released during oxide dissolution, thus preventing the formation of insoluble metal hydroxides or basic salts which result from hydrolysis reactions.[110] Förstner et al.[49] subdivided this phase into "easily reducible" and "moderately reducible", based on the ease of extraction of iron oxides. The "easily reducible" phase, incorporating metals occluded with manganese oxides and partly amorphous iron oxyhydrates, was extracted by NH$_2$OH.HCl coupled to HNO$_3$ (pH 2.0).[110] Extraction of the "moderately reducible" phase, consisting of metals occluded by poorly crystallized iron oxyhydrates, was by a mixture of ammonium oxalate ([CO$_2$NH$_4$]$_2$) and oxalic acid ([CO$_2$H]$_2$) at a pH value of 3.0.[111]

4. "Oxidizable" phase. Incorporating metals bound by organic matter and sulfides. Hot hydrogen peroxide (H$_2$O$_2$) in an HNO$_3$ medium is generally used to oxidize organic matter and extract sulfide minerals from sludges (Table 4). Oxidation by this reagent is extensive but incomplete, with the remaining organic matter consisting of paraffin-like material and resistant structural (nonhumidified) organic-matter residues.[40] Stronger oxidizing solutions may, however, seriously attack silicate material.[40] The choice of H$_2$O$_2$ thus represents a compromise between complete oxidation and alteration of silicate material. Steinhilber and Boswell[92] preferred to use sodium hypochlorite (NaOCl) to extract zinc occluded in organic matter. According to Lavkulich and Wiens,[112] NaOCl is more effective than H$_2$O$_2$ in oxidizing organic matter and less destructive of oxides and silicates. Using NaOCl also circumvents the problems of excessive frothing, oxalate formation, and manganese-oxide destruction associated with the use of H$_2$O$_2$.[112] Whether H$_2$O$_2$ or NaOCl is used as the initial reagent in the release of metals occluded by "oxidizable" material, extraction by NH$_4$OAc always follows,

with this latter reagent being effective in the removal of any liberated metal ions which become readsorbed.[97]

5. "Residual" phase. Consisting of metals held within the crystal lattices of primary and secondary minerals. Procedures for dissolving primary and secondary minerals involve digestion with mixtures of strong mineral acids (Table 4).

While the sequential extraction techniques employed by Förstner et al.[49] and Fraser and Lum[94] follow closely the order of extraction listed above, extraction of the "oxidizable" phase occurs between the "exchangeable" and "carbonate" steps in the technique used by Legret et al.[93] According to Meguellati et al.,[41] extraction of the "oxidizable" phase by H_2O_2 directly after the "exchangeable" phase allows for destruction of organic matter which has entrapped the mineral constituents of the matrix, thus facilitating the extraction of the following phases. Meguellati et al.[41] further demonstrated that the digestion of "oxidizable" matter by H_2O_2 had very little influence on the following "carbonate" and "reducible" phases. In contrast, however, Förstner et al.[49] and Lavkulich and Wiens[112] emphasized the possibility of reduction effects of the H_2O_2 treatment, principally through oxalate formation; hence the use of the H_2O_2 treatment after extraction of the "reducible" phase by Förstner et al.[49] and Fraser and Lum.[94] It is apparent, however, that when reagents such as NaOAc (primarily used for extraction of the "carbonate" phase) and $(CO_2NH_4)_2$ (primarily used for extraction of the "reducible" phase) are employed prior to the H_2O_2 procedure, they also extract significant amounts of organic matter and associated heavy metals.[41,113]

Sequential extraction techniques which have been used to fractionate heavy metals in sludge-amended soils are presented in Table 5. It is apparent that at least half of these procedures are modifications of the technique originally developed for use on sewage sludges by Stover et al.[26] The wider use of a method devised for a specific application may thus be useful in providing information regarding metal speciation in several related matrices on a comparative basis. Emmerich et al.[6] and Silviera and Sommers,[36] for example, compared the distribution of metals in sludge, soils, and sludge-soil mixtures by applying the same sequential extraction techniques to all three matrices.

2. Problems Associated with the Use of Sequential Extraction Techniques

The proliferation and diversity of reagents used to extract specific metal forms from wastewater matrices makes an intercomparison of results difficult. Tables 4 and 5 demonstrate the variation in interpretation between workers of the chemical forms of trace metals extracted by particular reagents, as illustrated by the incorporation of potassium nitrate into the extraction scheme of Schalscha et al.,[114] to extract water-soluble species, while in the majority of schemes, this reagent is employed to extract exchangeable forms. In addition, Steinhilber and Boswell[92] utilized EDTA to extract specifically adsorbed metal forms, while this reagent is primarily intended for the extraction of carbonate forms. Even when the same reagent is employed to extract a particular metal species, the rate and efficiency of leaching will be influenced by the type of sample, the size of particulates, and duration of extraction together with pH, temperature, and strength of extractant, and ratio of solid matter to volume of the extractant.[49,115] Despite proposals for the standardization of extraction procedures,[27,97] the effect of varying such parameters has received relatively little attention with respect to wastewater matrices. According to Salomons and Förstner,[97] however, a distinct reduction of the metal content in most extraction fractions is usually seen as sediment-grain-size increases. In addition, experiments by Calmano and Förstner[15] on polluted river sediments demonstrated that too high a solids content together with an increased buffer capacity may cause the system to overload. Such an effect was reflected by rising pH values in a time-dependent test with $NH_2OH \cdot HCl$ buffer (initial pH of 2.0) and ammonium oxalate-oxalic acid (pH 3.0) solutions.

Table 5
REAGENTS UTILIZED IN THE SEQUENTIAL EXTRACTION OF HEAVY METALS FROM SEWAGE SLUDGE-AMENDED SOILS

Designated chemical form	Cd, Cu, Ni, Pb, Zn (Ref. 39)	Cd, Cu, Pb, Zn (Ref. 36)	Cd, Cu, Ni, Zn (Ref. 6)	Cr, Cu, Mn, Ni, Zn (Ref. 114)	Cd, Cu, Ni, Pb, Zn (Ref. 134)	Cd, Cr, Cu, Ni, Pb, Zn (Ref. 141)	Cd, Cu, Ni, Zn (Ref. 136)
Soluble	H_2O	H_2O	(Footnote[a])				
Exchangeable	KNO_3	KNO_3	KNO_3	KNO_3	KNO_3	KNO_3	$MgCl_2$ (Footnote[b])
Adsorbed			Ion-exchange H_2O		Deionized H_2O	Deionized H_2O	
Organically bound			NaOH	$Na_4P_2O_7$	NaOH	NaOH	H_2O_2-HNO_3 and NH_4OAc
"Available"	DTPA	DPTA					
Carbonate							NaOAc
Sulfide			EDTA HNO_3 (Footnote[c])	EDTA HNO_3	EDTA HNO_3 (Footnote[c])	EDTA HNO_3 (Footnote[c])	
Oxide bound							$NH_2OH \cdot HCl$ and HOAc
Residual		HNO_3	HNO_3 (Footnote[c])			HNO_3 (Footnote[c])	HF-$HClO_4$ (Footnote[c])

[a] Soluble metals separated from saturation extract cake.
[b] Correct order of extraction scheme is exchangeable, carbonate, oxide-bound, organically bound, residual.
[c] Digestion procedures.
[d] Coprecipitated with or occluded in iron and manganese oxides.

Drying of a sludge sample in preparation for sequential extraction is almost a standard procedure, since this pretreatment gives improved homogeneity, stability, and ease of handling.[30,81,91] Investigations by Silviera and Sommers,[36] Oake et al.,[88] Rudd et al.,[30] and Lake[81] suggest, however, that the response of dried sludge samples is frequently different to that of liquid forms. In particular, air drying has consistently been found to cause a marked reduction in the concentrations of various heavy metals extracted by KNO_3, a reagent employed to remove the most labile soluble/exchangeable fraction.[30,81,88] Since the water content of sludge as disposed to land usually exceeds 65% even after dewatering treatment,[88] it may be more realistic to use liquid sludge for experimentation. This has its attendant problems, however, such as reduced homogeneity of samples[30,81] and the gradual alteration of sludge characteristics on storage.[88] Oake et al.,[88] for example, observed a transfer of approximately 20% of lead and zinc from the $Na_4P_2O_7$ (organically bound) extract to the EDTA (carbonate) and HNO_3 (sulfide) extracts following storage of liquid sludges for 1 month, suggesting a possible loss of specific binding sites on the organic material.

Unfortunately, sequential extraction procedures represent only arbitrary divisions between different metal forms, since reagents used are not entirely selective for one particular form[8,115] and rarely are they complete in their solubilization of a target phase.[81] Reagents such as $Na_4P_2O_7$, primarily intended for the extraction of organic metal fractions, have been shown to extract 79 and 18% of model precipitates of lead carbonate and lead sulfide, respectively.[26] Rudd and Lester[116] similarly found that $Na_4P_2O_7$ is able to extract carbonate (44%) and sulfide (14%) forms of lead in addition to substantial concentrations of lead phosphate (75%) and hydroxide (55%) precipitates. Extraction of these forms by $Na_4P_2O_7$ may occur through the formation of lead pyrophosphate ($Pb_2P_2O_7 \cdot H_2O$), which is soluble in alkaline solution.[117] The specificity of HNO_3 for sulfide forms appears to be metal dependent; in excess of 80% of model precipitates of cadmium, nickel, and lead sulfides having been extracted by this reagent, but less than 20% with respect to copper and zinc.[116] Similarly, EDTA extracted more than 90% of model precipitates of copper, lead, and zinc carbonates but only 68% of cadmium carbonate.[26] The nonselective nature of EDTA, which is primarily intended for the removal of carbonate metal forms, is emphasized by its ability to also extract substantial concentrations of phosphate and hydroxide precipitates of cadmium, copper, nickel, lead, and zinc in addition to nickel and lead sulfides.[116]

The specificity of reagents such as NaOAc, $NH_2OH \cdot HCl$, H_2O_2, NH_4OAc, $BaCl_2$, and $MgCl_2$ towards particular metal phases has been investigated by Meguellati et al.[41] and Rapin and Förstner;[118] the former used synthetic phases of montmorillonite, carbonate, $Fe(OH)_3$, MnO_2, and humic acid, while the latter used actual, well defined mineral phases. These investigations revealed that (1) Na_4OAc, primarily intended for extraction of the "exchangeable" phase, partially dissolved carbonate forms of cadmium, lead, and manganese;[118] (2) $MgCl_2$ and $BaCl_2$, also used to extract the "exchangeable" phase, had the ability to attack copper and lead associated with "carbonate", "oxide", and "organic" phases;[41] and (3) NaOAc, primarily intended for extraction of carbonate forms, was found to extract 90% of the lead associated with organic (humic) material[41] in addition to 80% of the cadmium associated with iron and manganese oxides.[118]

Recent investigations by Rudd and Lester[116] suggested that, in the majority of cases, measurement of reagent specificity by discrete extraction of a particular metal form does not predict the behavior of this extractant under sequential application. For example, Rudd and Lester[116] found that although under discrete extraction the majority of the metal carbonate precipitates used were extracted by EDTA, the transition to sequential extraction resulted in poor specificity of EDTA for carbonates of nickel and lead in particular. Such differences may be caused either by concentration factors (i.e., when there is less metal left to extract, proportionally less will be dissolved) or by the modifying effects of preceeding extractants on the compound characteristics. Whatever the reason, the observation of poor predictability seems particularly significant because several previous investigations of specificity using model compounds[26,41] have only applied the reagent singly and assumed a direct extrapolation in their sequential application.

It is thus apparent that the distribution of a given metal between various fractions can only be considered as operationally defined by the method of extraction. Before metal concentrations determined in a particular fraction can be ascribed with reasonable certainty to defined geochemical forms, it will be necessary to examine present techniques in greater detail and develop more selective extraction reagents. In the choice of extractants, less powerful leaching solutions will probably be more selective for specific fractions than more stringent reagents which may attack other forms, although the overall efficiency of extraction may be lower.[8,115]

B. Heavy-Metal Distributions in Sewage Sludges and Related Matrices

Despite the shortcomings of sequential extraction techniques, they have provided a wealth of information on the distribution of heavy metals in sewage sludges and related matrices. In view of the recognized nonspecific nature of such methods, however, metal species are hereafter defined principally by their reagent of extraction rather than being ascribed to precise physicochemical forms. Such an approach also facilitates comparison of the diverse results obtained by numerous workers since it is not unreasonable to assume that chemically similar forms of an individual metal will be extracted from various sludges by a particular reagent.

The forms of heavy metals present in sewage sludges vary widely according to the nature of the individual metal, the characteristics of the wastewater treated, and the sludge treatment employed.[33,115] Using a similar extraction scheme, Oake et al.,[88] Rudd et al.,[30] and Lake[81] found, however, that speciation profiles obtained for cadmium, copper, nickel, lead, and zinc in raw and activated sludges originating from U.K. sewage treatment works were essentially independent of sludge type, showing greater dependence on the nature of the individual metals. This would seem to suggest that major mechanisms of metal retention, such as ion-exchange, adsorption, complexation, and precipitation, were common to both types of matrix.[30,81] According to sequential chemical extraction, the major forms of metal present in the sludges were HNO_3-extractable (sulfide) for copper, EDTA-extractable (carbonate) for cadmium and nickel, and $Na_4P_2O_7$-extractable (organically bound) for lead and zinc.[30,81,88] In agreement with these findings, $Na_4P_2O_7$-extractable forms of lead and zinc were also found to predominate in raw primary and mixed primary (i.e. cosettled primary and surplus activated) sludges investigated in a separate study by Lake et al.[91] In contrast, however, nickel occurred mainly in KNO_3-extractable (soluble/exchangeable) and "residual" forms, while $Na_4P_2O_7$ (organically bound) and EDTA (carbonate) fractions predominated for cadmium and copper, respectively.[91]

The predominance of lead and zinc in $Na_4P_2O_7$-extractable forms would seem to support the hypothesis put forward by Gould and Genetelli[22] that organometallic complexation may be a major mechanism responsible for the association of heavy metals with insoluble sludge solids. Lead, in particular, is known to have a high affinity for organic ligands in sludges[119,120] and also soils.[85,86,121] In each of the sequential extraction studies,[30,81,88,91] zinc was the only metal for which the KF-extractable (adsorbed) fraction was significant, accounting for up to 20% of the total zinc content of some raw sludges.[91] This concurs with the suggestion made by Adams and Sanders[51] that adsorption/desorption equilibria, including cation-exchange processes, are important for the retention of zinc in sewage sludges. In addition, nickel was the only metal for which KNO_3-extractable (soluble/exchangeable) forms were significant, averaging 13% of the total nickel content of the raw and activated sludges studied by Lake[81] and approximately 30% of those studied by Oake et al.[88] and Lake et al.[91] The predominance of cadmium in the "organically bound" fraction[91] would seem to support the work of Riffaldi et al.,[122] who found organic complexing sites to play an important role in the binding of cadmium to sewage sludge solids. The predominance of copper in raw sludges in either the EDTA (carbonate) fraction[91] or the HNO_3 (sulfide) fraction[30,81,88] is, however, contrary to its known affinity for organic ligands.[78,119,120] The persistence of "sulfide" forms of copper in activated sludges[30,81,88] would appear to be particularly anomalous as sulfides are generally not formed under the aerobic conditions which prevail in the activated-sludge system.[123] As mentioned previously, however, HNO_3 has been shown to extract less than 20% of model precipitates of copper sulfide.[116] In addition, some doubt has been expressed as to the efficacy of $Na_4P_2O_7$ as an extractant of copper organic species.[102] Therefore, the possibility exists that the HNO_3 fraction may contain some organic forms of copper. According to Förstner et al.,[49] copper is predominantly present in activated sludges in ammonium-oxalate-extractable (occluded with iron oxide) forms.

Guidelines for sludge application to agricultural land in the U.K. are restrictive for raw sludges,[89] and stabilization of sludge prior to such disposal is likely to become mandatory throughout the EEC.[124] The most widely applied sludge-stabilization process in the U.K. is mesophilic anaerobic digestion,[89] to which approximately 60% of the total sludge produced is subjected.[125] Lake et al.[91] investigated the transitions which occur in the metal distributions of raw sludges following anaerobic treatment in laboratory-scale digesters. It was observed that, in contrast to cadmium and lead, concentrations of copper, nickel, and zinc in the most labile KNO_3-extractable (soluble/exchangeable) fraction were markedly reduced as a consequence of digestion,[91] as illustrated in Figure 5. This is consistent with the increase in water solubility of the former two metals and reduction of the latter three observed by Bloomfield and Pruden[24] following anaerobic incubation of digested sludge for 3 months. The greater significance of "soluble" and/or "exchangeable" forms of copper, nickel, and zinc in aerobic as opposed to anaerobic sludge matrices has also been reported by Legret et al.,[93] Oake et al.,[88] Rudd et al.,[30] and Lake.[81] According to Mosey,[126] this reduction in heavy-metal solubility may be attributed to the formation of heavy-metal sulfide precipitates. Although markedly reduced as a consequence of digestion, a significant proportion of the total nickel content of the anaerobically digested sludges studied by Lake et al.[91] remained in the KNO_3 (soluble/exchangeable) fraction, averaging 19%. Stover et al.,[26] Emmerich et al.,[6] and Oake et al.[88] also observed this fraction to be of greater significance to nickel than to any of the other metals they studied, constituting up to 27% of the total nickel content of some anaerobically digested sludges.[26]

In general, however, Lake et al.[91] found that the overall effect of anaerobic digestion was to cause a shift away from the more easily extractable forms which predominated in the raw sludges towards less readily extractable "precipitated" forms (Figure 5). This was particularly evident with respect to cadmium and lead, which predominated in the digested sludges in EDTA-extractable (carbonate) forms as opposed to the more readily extractable $Na_4P_2O_7$ (organically bound) forms observed in raw sludges. In addition, HNO_3-extractable (sulfide) forms of copper were preeminent in the digested sludges as opposed to the less stable EDTA-extractable (carbonate) forms which predominated in the raw sludges. Zinc, however, remained essentially in the $Na_4P_2O_7$ (organically bound) fraction, while nickel was distributed evenly among all but the KF (adsorbed) fraction.[91] Stover et al.[26] similarly reported cadmium and lead to be predominantly present in anaerobically digested sludge in EDTA (carbonate) forms, with copper and zinc existing mainly in HNO_3 (sulfide) and $Na_4P_2O_7$ (organically bound) fractions, respectively, as detailed in Table 6. In contrast, however, nickel predominated in the EDTA-extractable (carbonate) fraction.[26] A similar extraction procedure applied by Emmerich et al.[6] to digested sludge indicated that, in accordance with the work of Stover et al.,[26] cadmium and nickel occurred predominantly in the EDTA-extractable (carbonate) fraction. The predominance of zinc in this fraction, however, disagrees with the findings of Stover et al.[26] and Lake et al.[91] In addition, Emmerich et al.[6] reported the major forms of

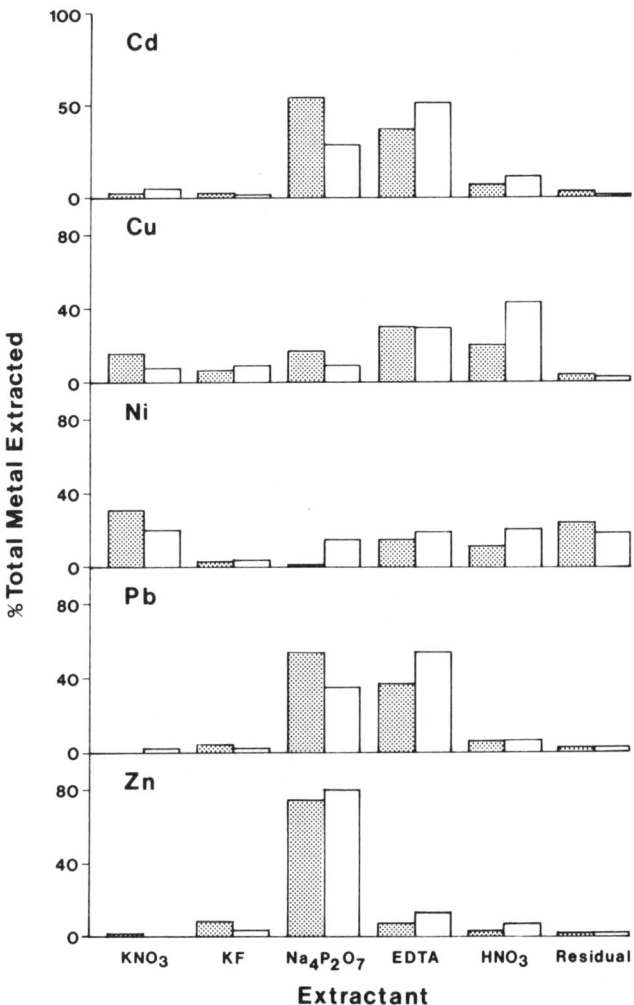

FIGURE 5. Comparison of metal distributions in raw (▨) and digested (□) primary sludge resulting from sequential chemical extraction. (From Lake, D. L., Kirk, P. W. W., and Lester, J. N., *Water Pollut. Control*, 84, 549, 1985. With permission.)

Table 6
DISTRIBUTION OF METAL FORMS (IN % TOTAL METAL) IN DIGESTED SLUDGE BY SEQUENTIAL EXTRACTION[26]

Metal	Extractant[a]				
	KNO_3	KF	$Na_4P_2O_7$	EDTA	HNO_3
Cd	0	0	14.8	48.8	17.5
Cu	6.4	10.4	10.4	22.5	35.1
Ni	13.9	8.3	14.2	32.4	6.8
Pb	0	8.8	29.1	61.4	4.4
Zn	0.3	0.4	50.3	18.2	9.3

[a] KNO_3, exchangeable; KF, adsorbed; $Na_4P_2O_7$, organically bound; EDTA, carbonate; HNO_3, sulfide.

copper extracted to be in the order NaOH (organically bound) > EDTA (carbonate) > HNO_3 (sulfide/residual), with in excess of 60% of the sludge copper being found in the "organically bound" form. This sequence is the reverse of that reported for copper by Stover et al.[26] and Lake et al.;[91] this is a possible reflection of the greater efficacy of NaOH relative to $Na_4P_2O_7$ as an extractant of copper-organic species. Also in apparent contradiction to the results of Stover et al.,[26] Legret et al.[93] demonstrated that copper in digested sewage sludges was largely associated with the "organic" phase, which represented 50% of the total copper content. However, closer inspection of the form designated as "organic" reveals the inclusion of "sulfides" as an H_2O_2-extractable "oxidizable" phase (Table 4), suggesting that the ranking of copper species by Stover et al.[26] and Legret et al.[93] may not be as dissimilar as they initially appear.

Despite some apparent inconsistencies, results obtained from the sequential extraction of heavy metals in anaerobically digested sludges suggest the predominance of inorganically precipitated forms. This is not unexpected since it is known that all heavy metals, with the exception of chromium, form extremely insoluble sulfide salts under the reducing conditions of digestion;[127] this is via the reduction of sulfur compounds including sulfate and the amino acids cysteine and methionine.[126] In addition, certain metals have been shown to be precipitated as sparingly soluble carbonate salts provided that the digester pH is sufficiently high, with cadmium and zinc being bound as carbonates at pH values exceeding pH 7.2[128] and pH 7.7,[126] respectively.

In support of the predominance of metals such as zinc in $Na_4P_2O_7$ extractable (organically bound) forms, there is mounting evidence to suggest that organometallic interactions are important mechanisms of metal retention in digested sludge.[9,22] Tan et al.[129] studied complex formation between zinc and alkali-extracted organic matter from anaerobically digested sewage sludge and concluded, on the basis of infrared spectra, that hydroxyl and carboxyl functional groups were involved in zinc binding. Involvement of thiol functional groups[130] and amide groups of peptide bonds in residual proteins[131] has also been suggested however. In addition, Hayes and Theis[132] reported zinc to be present in anaerobically digested sludge in a predominantly intracellular form, suggesting that zinc may be actively taken up by the microbial population within digesters.

The data obtained from sequential chemical extractions largely confirm those of single chemical extractions with respect to the stabilization of metal forms in sewage sludge as a result of anaerobic digestion. Since metal mobility and bioavailability are considered to decrease approximately in the order of an extraction sequence, it is likely that heavy metals present in sewage sludges disposed of to land would be of less immediate concern following anaerobic digestion, as demonstrated by Petruzzelli et al.[133] Consideration of the influence of total metal concentration and sludge characteristics on metal speciation, particularly with respect to cadmium and nickel, is necessary, however, in the safe disposal of digested sewage sludge.[91]

Sequential chemical-extraction techniques can further elucidate the transitions in metal form which occur following sludge disposal to land. Using such techniques, it has been found that concentrations of metals extracted from sludge-amended soils in the most labile "soluble", "exchangeable", and "adsorbed" fractions, although often marginally higher than those extracted from respective unamended soils,[6,36] are generally very low. Emmerich et al.,[6] using a modified version of the technique developed by Stover et al.,[26] observed that <3% of the total cadmium, copper, nickel, and zinc content of sludge-amended loams was extracted by KNO_3 ("exchangeable" fraction) and "ion-exchange water" ("adsorbed" fraction). Sposito et al.[134] similarly extracted 1.1 to 3.7% of the total cadmium, copper, nickel, lead, and zinc of amended field soils using the same reagents. Such results are consistent with those reported for KNO_3 (soluble/exchangeable) and KF (adsorbed) forms of cadmium, lead, and zinc in sludges;[26] H_2O-extractable (soluble) plus KNO_3 (exchangable) forms of cadmium, copper, lead, and zinc in silt loam soils incubated with digested sludge,[36] and KNO_3 (soluble) and NaF (exchangeable) forms of copper and zinc in Chilean inceptisols irrigated with raw sewage.[114] The small percentages of metals present in sludge-amended soils in these easily solubilized fractions may signify their low immediate availability for plant uptake, since bioavailability is reported to decrease in the order of an extraction sequence.[135,136] However, according to Silviera and Sommers,[36] Schalscha et al.,[114] and Sposito et al.,[134] the lack of "soluble" and "exchangeable" forms indicates a low leaching potential of metals in sludge-amended soils rather than a low availability to crops. Heavy metals added to soil in sewage sludges have often been reported to exhibit little mobility, even years after introduction, with most of the added metal remaining in the surface horizon.[6,71,137-139]

It is evident, however, that the application of sewage sludge to soils generally causes a shift in solid-phase metal forms away from those extractable with severe reagents, such as HNO_3-extractable (sulfide/residual) forms, to those extractable with milder reagents.[6,114,134,140] This is particularly apparent with increasing sludge-application rate.[134] Using similar extraction techniques (Table 5), Emmerich et al.[6] Sposito et al.,[134] and Chang et al.[141] found that heavy metals were predominantly present in unamended soils in either HNO_3-extractable (sulfide/residual) forms (chromium, copper, nickel, and zinc) or EDTA-extractable (carbonate) forms (cadmium and lead). Following sludge application, however, large reductions in "sulfide/residual" forms of essentially all metals studied were simultaneously accompanied by significant increases in NaOH (organically bound) and EDTA (carbonate) forms.[6,134,141] While such changes did not appear to significantly alter the orders of predominant forms of cadmium, nickel, and lead from those found in unamended soils, zinc predominated in sludge-amended soils in the more easily extractable EDTA (carbonate) fraction.[6,134] Similarly, NaOH (organically bound) forms became most significant with respect to copper, representing up to 60% of the total copper content of sludge-soil mixtures studied by Sposito et al.[134] This is consistent with the known affinity of copper for organic ligands.[11,21,87] Such effects suggest that the application of sewage sludge to soils could provide metals in labile chemical forms that might be potentially more bioavailable than those in nonamended soils. Indeed, marked increases in the amounts of total cadmium, copper, lead, and zinc extractable by DTPA (believed to estimate "available" metal species) were observed in two silt loam soils 28 days after sewage sludge had been applied.[36] Using a similar scheme, Petruzzelli et al.[39] reported that DTPA-extractable forms of copper, nickel, lead, and zinc in an acidic sandy loam soil reached maximum values 30 days after the addition of sludge but decreased at the time of crop harvest. The apparent increase in DTPA-extractable metals from sludge-soil mixtures with time was ascribed to dissolution of metal precipitates, such as carbonates, hydroxides, and phosphates, through changes in pH or gas composition of the soil resulting from microbial activity; oxidation of metal sulfides to sulfates by autotrophic sulfur-oxidizing bacteria; and microbial release of metals complexed with sludge organic matter.

It has been suggested by Emmerich et al.,[6] however, that over longer time periods there may be a tendency for metal forms in sludges applied to soils to stabilize and gradually revert to the "residual" form, possibly via fixation with inorganic sludge components such as iron, aluminum, or manganese oxides.[51,92] This phenomenon has been ascribed by Steinhilber and Boswell[92] and Schaumberg et al.[142] to the progressive loss of soluble-phase functional groups for metal binding, as demonstrated by infrared spectroscopy.

REFERENCES

1. American Public Health Association, *Standard Methods for the Examination of Water and Wastewater*, 14th ed., A.P.H.A., Washington, D. C., 1975, 143.
2. **Sterritt, R. M. and Lester, J. N.**, Atomic absorption spectrophometric analysis of the metal content of wastewater samples, *Environ. Technol. Lett.*, 1, 402, 1980.
3. Department of the Environment, A Survey of Multi-Element and Related Methods of Analysis of Waters, Sediments and Other Related Materials of Interest to the Water Industry, Her Majesty's Stationery Office, London, 1981, 1.
4. **Kempton, S., Sterritt, R. M., and Lester, J. N.**, Atomic absorption spectrophotometric determination of antimony, arsenic, bismuth, tellurium, thallium and vandadium in sewage sludge, *Talanta*, 29, 675, 1982.
5. **Norton, R. L. and Orpwood, J. R.**, The application of inductively coupled plasma spectrometry to the analysis of metals in environmental samples, in *Proc. Int. Conf. Environmental Contamination*, CEP Consultants Ltd., Edinburgh, 1984, 172.
6. **Emmerich, W. E., Lund, L. J., Page, A. L., an Chang, A. C.**, Solid phase forms of heavy metals in sewage sludge-treated soil, *J. Environ. Qual.*, 11, 178, 1982.
7. **Lester, J. N., Sterritt, R. M., and Kirk, P. W. W.**, Significance and behaviour of heavy metals in wastewater treatment processes. II. Sludge treatment and disposal, *Sci. Total Environ.*, 30, 45, 1983.
8. **Lake, D. L., Kirk, P. W. W., and Lester, J. N.**, Fractionation, characterization, and speciation of heavy metals in sewage sludge and sludge-amended soils: a review, *J. Environ. Qual.*, 13, 175, 1984.
9. **Gould, M. S. and Genetelli, E. J.**, Heavy metal distribution in anaerobically digested sludges, in *Proc. 30th Industrial Waste Conf. Purdue University*, May 1975, Ann Arbor Science, Ann Arbor, Mich., 1977, 689.
10. **Bell, C. F.**, *Principles and Applications of Metal Chelation*, Oxford University Press, Oxford, U.K., 1977, 1.
11. **Stevenson, F. J. and Ardakani, M. S.**, Organic matter reactions involving micronutrients in soils, in *Micronutrients in Agriculture*, Mortvedt, J. J., Giordano, P. M., and Lindsay, W. L., Eds., Soil Science Society of America, Madison, Wisc., 1972, chap. 5.
12. **Worthington, P.**, A closer look at metals in sewage sludge, in *Proc. 1st European Symp. Treatment and Use of Sewage Sludge*, Alexander, D. and Ott, H., Eds., D. Reidel, Dordrecht, the Netherlands, 1979, 144.
13. **Schnitzer, M.**, Reactions between fulvic acid, a soil humic compound and inorganic soil constituents, *Soil Sci. Soc. Am. Proc.*, 33, 75, 1969.
14. **Ellis, B. G. and Knezek, B. D.**, Adsorption reactions of micronutrients in soils, in *Micronutrients in Agriculture*, Mortvedt, J. J., Giordano, P. M., and Lindsay, W. L., Eds., Soil Science Society of America, Madison, Wisc., 1972, chap. 4.
15. **Calmano, W. and Förstner, U.**, Chemical extraction of heavy metals in polluted river sediments in central Europe, *Sci. Total Environ.*, 28, 77, 1983.
16. **Doyle, P. J., Lester, J. N., and Perry, R.**, Survey of Literature and Experience on the Disposal of Sewage Sludge on Land, Report DGR/480/60, U.K. Department of the Environment, London, 1978.
17. **Wolf, A., Bunzl, K., Dietl, F., and Schmidt, W. F.**, Effect of Ca^{2+}-ions on the adsorption of Pb^{2+}, Cu^{2+}, Cd^{2+}, and Zn^{2+} by humic substances, *Chemosphere*, 5, 207, 1977.
18. **McLaren, R. G., Swift, R. S., and Williams, J. G.**, The adsorption of copper by soil materials at low equilibrium concentrations, *J. Soil Sci.*, 32, 247, 1981.
19. **Wildung, R. E., Garland, T. R., and Drucker, H.**, Nickel complexes with microbial metabolites — mobility and speciation in soils, in *Chemical Modeling in Aqueous Systems*, Jenne, E. A., Ed., American Chemical Society Symp. Ser. No. 93, American Chemical Society, Washington D.C., 1979, 181.
20. **Parker, S. P.**, *Dictionary of Scientific and Technical Terms*, 3rd ed., McGraw-Hill, New York, 1984, 1260.
21. **McLaren, R. G. and Crawford, D. V.**, Studies on soil copper. I. The fractionation of copper in soils, *J. Soil Sci.*, 172, 1973.
22. **Gould, M. S. and Genetelli, E. J.**, Heavy metal complexation behavior in anaerobically digested sludges, *Water Res.*, 12, 505, 1978.
23. **Berrow, M. L. and Webber, J.**, Trace elements in sewage sludges, *J. Sci. Food Agric.*, 23, 93, 1972.
24. **Bloomfield, C. and Pruden, G.**, The effects of aerobic and anaerobic incubation on the extractabilities of heavy metal in digested sewage sludge, *Environ. Pollut. (Ser. B)*, 8, 217, 1975.
25. **Bradford, G. R., Page, A. L., Lund, L. J., and Olmstead, W.**, Trace element concentrations of sewage treatment plant effluents and sludges: their interactions with soils and uptake by plants, *J. Environ. Qual.*, 4, 123, 1975.
26. **Stover, R. C., Sommers, L. E., and Silviera, D. J.**, Evaluation of metals in wastewater sludge, *J. Water Pollut. Control Fed.*, 48, 2165, 1976.

27. **Neuhauser, E. F. and Hartstein, R.,** Efficiencies of extractants used in analyses of heavy metals in sludges, *J. Environ. Qual.,* 9, 21, 1980.
28. **Bloomfield, C., and McGareth, S. P.,** A comparison of the extractabilities of Zn, Cu, Ni and Cu from sewage sludges prepared by treating raw sewage with the metal salts before and after anaerobic digestion, *Environ. Pollut. (Ser. B),* 193, 1982.
29. **Adams, T. McM.,** The effect of pH on the uptake of Zn, Cu and Ni from chloride solutions by an uncontaminated sewage sludge, *Environ. Pollut. (Ser. B),* 9, 151, 1985.
30. **Rudd, T., Lake, D. L., Kirk, P. W. W., and Lester, J. N.,** Chemical Fractionation of Heavy Metals in Sewage Sludges, 2nd Report to Water Research Centre, Medmenham, U.K., 1985.
31. **Jenkins, S. H. and Cooper, J. S.,** The solubility of heavy metal hydroxides in water, sewage and sewage sludge. III. The solubility of metals present in digested sewage sludge, *Int. J. Air Water Pollut.,* 8, 695, 1964.
32. **Lagerwerff, J. V., Biersdorf, G. T., and Brower, D. L.,** Retention of metals in sewage sludge. I. Constituent heavy metals, *J. Environ. Qual.,* 5, 19, 1976.
33. **Kirk, P. W. W., Lake, D. L., Lester, J. N., Rudd, T., and Sterritt, R. M.,** Metal speciation in sewage, sewage sludge and sludge-amended soil and sea water — a review, Tech. Rep. TR 226, Water Research Centre, Medmenham, U.K., 1985, 1.
34. **Latterell, J. J., Dowdy, R. H., and Larson, W. E.,** Correlation of extractable metals and metal uptake of snap beans grown on soils amended with sewage sludge, *J. Environ. Qual.,* 7, 435, 1978.
35. **Carlton-Smith, C. H. and Davis, R. D.,** An interlaboratory comparison of metal determinations in sludge-treated soil, *Water Pollut. Control,* 82, 544, 1983.
36. **Silviera, D. J. and Sommers, L. E.,** Extractibility of copper, zinc, cadmium and lead in soils incubated with sewage sludge, *J. Environ. Qual.,* 6, 47, 1977.
37. **Davis, R. D.,** Sludge disposal — keeping it safe, *Water Waste Treat.,* 27, 38, 1984.
38. **Matthews, P. J.,** Control of metal application rates from sewage sludge utilisation in agriculture, *Crit. Rev. Environ. Control.,* 14, 199, 1984.
39. **Petruzzelli, G., Lubrano, L., and Guidi, G.,** The effect of sewage sludge and composts on the extractability of heavy metals from soil, *Environ. Technol. Lett.,* 2, 449, 1981.
40. **Tessier, A., Campbell, P. G. C., and Bissom, M.,** Sequential extraction procedure for the speciation of particulate trace metals, *Anal. Chem.,* 51, 844, 1979.
41. **Meguellati, N., Robbe, D., Marchandise, P., and Astruc, M.,** A new chemical extraction procedure in the fractionation of heavy metals in sediments — interpretation, in *Proc. Int. Conf. Heavy Metals in the Environment,* Vol. 2, CEP Consultants Ltd., Edinburgh, 1983, 1090.
42. **Elsokarry, I. H. and Lagg, J.,** Distribution of different fractions of Cd, Pb, Zn and Cu in industry polluted and non-polluted soils of Odda Region, Norway, *Acta Agric. Scand.,* 28, 262, 1978.
43. **Kuo, S., Herlman, P. E., and Baker, S.,** Distribution and forms of Cu, Zn, Cd, Fe and Mn in soils near a copper smelter, *Soil Sci.,* 135, 101, 1983.
44. **Cavallaro, N. and McBride, M. B.,** Zinc and copper sorption and fixation by an acid soil clay: effect of selective dissolutions, *Soil Sci. Soc. Am. J.,* 48, 1050, 1984.
45. **Adams, F.,** Manganese, in *Methods of Soil Analysis,* Vol. 2., Black, C. A., Evans, D. D., White, J. L., Ensminger, L. E., and Clark, F. E., Eds., American Society of Agronomy, Madison, Wisc., 1965, 1011.
46. **Cox, F. R. and Kamprath, E. J.,** Micronutrient soil tests, in *Micronutrients in Agriculture,* Mortvedt, J. J., Giordano, P. M., and Lindsay, W. L., Eds., Soil Science of America, Madison, Wisc., 1972, chap. 13.
47. **Cottenie, A. and Verloo, M.,** Analytical diagnosis of soil pollution with heavy metals, *Fresenius Z. Anal. Chem.,* 317, 389, 1984.
48. **Sauerbeck, D. R. and Rietz, E.,** Soil-chemical evaluation of different extractants for heavy metals in soils, in *Environmental Effects of Organic and Inorganic Contaminants in Sewage Sludge,* Davis, R. D., Hucker, G., and L'Hermite, P., Eds., D. Reidel, Dordrecht, the Netherlands, 1983, 147.
49. **Förstner, U., Calmano, W., Conradt, K., Jaksch. H., Schimkus, C., and Schoer, J.,** Chemical speciation of heavy metals in solid waste materials (sewage sludge, mining wastes, dredged materials, polluted sediments) by sequential extraction, in *Proc. Int. Conf. Heavy Metals in the Environment,* CEP Consultants Ltd., Edinburgh, 1981, 698.
50. **Davis, R. D.,** Crop uptake of metals (cadmium, lead, mercury, copper, nickel, zinc and chromium) from sludge treated soil in its implication for soil fertility and for the human diet, in *Processing and Use of Sewage Sludge,* L'Hermite, P. and Ott, H., Eds., D. Reidel, Dordrecht, the Netherlands, 1984, 349.
51. **Adams, T., McM. and Sanders, J. R.,** The effects of pH on the release to solution of zinc, copper and nickel from metal-loaded sewage sludges, *Environ. Pollut. (Ser. B),* 8, 85, 1984.
52. **Wear, J. I. and Evans, C. E.,** Relationship of zinc uptake by corn and sorghum to soil zinc measured by three extractants, *Soil Sci. Soc. Am. Proc.,* 32, 543, 1968.
53. **Norvell, W. A. and Lindsay, W. L.,** Reactions of EDTA complexes of Fe, Zn, Mn, and Cu with soils, *Soil Sci. Soc. Am. Proc.,* 33, 86, 1969.

54. **Trierweiler, J. F. and Lindsay, W. L.**, EDTA — ammonium carbonate soil test for zinc, *Soil Sci. Soc. Am. Proc.*, 33, 49, 1969.
55. **Norvell, W. A. and Lindsay, W. L.**, Reactions of DPTA chelates of iron, zinc, copper, and manganese with soils, *Soil Sci. Soc. Am. Proc.*, 36, 778, 1972.
56. **Misra, S. G. and Pande, P.**, Evaluation of a suitable extractant for available nickel in soils, *Plant Soil*, 41, 697, 1974.
57. **Lindsay, W. L. and Norvell, W. A.**, Development of a DTPA soil test for zinc, iron, manganese, and copper, *Soil Sci. Soc. Am. J.*, 42, 421, 1978.
58. **Tiwari, R. C. and Mohankumar, B.**, A suitable extractant for assessing plant-available copper in different soils (peaty, red and alluvial), *Plant Soil*, 68, 131, 1982.
59. **Viets, F. G., Jr.**, Chemistry and availability of micronutrients, *J. Agric. Food Chem.*, 10, 174, 1962.
60. **Davis, R. D. and Carlton-Smith, C. H.**, An inter-laboratory comparison of metal determinations in sewage sludges and soil, *Water Pollut. Control*, 82, 290, 1983.
61. **Ellis, R. H. and Alloway, B. J.**, Factors affecting the availability of cadmium, lead and nickel in soils amended with sewage sludge, in *Proc. Int. Conf. Heavy Metals in the Environment*, Vol. I, CEP Consultants Ltd., Edinburgh, 1983, 358.
62. **Wollan, E. and Beckett, P. H. T.**, Changes in the extractability of heavy metals on the interaction of sewage sludge with soil, *Environ. Pollut. (Ser. B)*, 20, 215, 1979.
63. **Gaynor, J. D. and Halstead, R. L.**, Chemical and plant extractability of metals and plant growth on soil amended with sludge, *Can. J. Soil Sci.*, 56, 1, 1976.
64. **Kelling, K. A., Keeny, D. R., Walsh, L. M., and Ryan, J. A.**, A field study of the agricultural use of sewage sludge. III. Effect on uptake and extractability of sludge-borne metals, *J. Environ. Qual.*, 6, 352, 1977.
65. **Street, J. J., Lindsay, W. L., and Sabey, B. R.**, Solubility and plant uptake of cadmium in soils amended with cadmium and sewage sludge, *J. Environ. Qual.*, 6, 72, 1977.
66. **Schauer, P. S., Wright, W. R., and Pelchat, J.**, Sludge borne heavy metal availability and uptake by vegetable crops under field conditions, *J. Environ. Qual.*, 9, 69, 1980.
67. **Beckett, P. H. T., Warr, E., and Brindley, P.**, Changes in the extractabilities of the heavy metals in water-logged sludge-treated soils, *Water Pollut. Control*, 82, 107, 1983.
68. **Mitchell, G. A., Bingham, F. T., and Page, A. L.**, Yield and metal composition of lettuce and wheat grown on soils amended with sewage sludge enriched with cadmium, copper, nickel and zinc, *J. Environ. Qual.*, 7, 165, 1978.
69. **Beckett, P H. T., Warr, E., and Davis, R. D.**, Cu and Zn in soils treated with sewage sludge: their 'extractability' to reagents compared with their 'availability' to plants, *Plant Soil*, 70, 3, 1983.
70. **Steinhilber, P. M.**, Fate of Sewage Sludge Derived Zn Relative to Soil Factors and Plant Utilization, Ph.D. dissertation, University of Georgia, University Microfilms, Ann Arbor, Mich., 1981.
71. **Williams, D. E., Vlamis, J., Pukite, A. H., and Corey, J. E.**, Trace element accumulation, movement and distribution in the soil profile from massive application of sewage sludges, *Soil Sci.*, 129, 119, 1980.
72. **Leeper, G. W.**, Reactions of Heavy Metals with Soil, with Special Regard for Their Application in Sewage Wastes, Department of the Army, Army Corps of Engineers, Contract No. DACW/73-73-C-0026, 1972, 26.
73. **Cottenie, A.**, Mobility of heavy metals in sludge amended soil, in *Characterization, Treatment and Use of Sewage Sludge*, L'Hermite, P. and Ott, H., Eds., D. Reidel, Dordrecht, the Netherlands, 1981, 251.
74. **Soon, K. Y.**, Solubility and sorption of cadmium in soils amended with sewage sludge, *J. Soil Sci.*, 32, 85, 1981.
75. **Gerritse, R. G., Vriesema, R., Dalenberg, J. W., and De Roos, H. P.**, Effect of sewage sludge on trace element mobility in soils, *J. Environ. Qual.*, 11(3), 359, 1982.
76. **Sanders, J. R.**, The effect of pH on the total and free ionic concentrations of manganese, zinc and cobalt in soil solutions, *J. Soil Sci.*, 34, 315, 1983.
77. **Dhaese, A. and Cottenie, A.**, Contents of heavy metals in sludges and their environmental significance, in *Proc. 1st European Symp. Treatment and Use of Sewage Sludge*, Alexandre, J. and Ott, H., Eds., D. Reidel, Dordrecht, 1979, 364.
78. **Kiekens, L. and Cottenie, A.**, Report of the results of the interlaboratory comparison: determination of the mobility of heavy metals in soils, in *Processing and Use of Sewage Sludge*, L'Hermite, P. and Ott, H., Eds., D. Reidel, Dordrecht, the Netherlands, 1984, 140.
79. **Cottenie, A., Kiekens, L., and Van Landschoot, G.**, Problems of mobility and predictability of heavy metal uptake by plants, in *Processing and Use of Sewage Sludge*, L'Hermite, P. and Ott, H., Eds., D. Reidel, Dordrecht, the Netherlands, 1984, 124.
80. **Adams, T. McM.**, Chemistry of zinc, copper and nickel from sewage sludge and sludge-treated soils, in *Proc. Int. Conf. Heavy Metals in the Environment*, Vol. 1, CEP Consultants Ltd., Edinburgh, 1983, 479.
81. **Lake, D. L.**, Characterization and Fractionation of Heavy Metals in Sewage Sludge, Ph.D. thesis, University of London, 1986.

82. **Adams, T. McM., and Sanders, J. R.**, The effect of pH, soil type and incubation on the release to solution of zinc, copper and nickel from 3 soils treated with metal-loaded sewage sludges, in *Proc. Int. Conf. Environmental Contamination*, CEP Consultants Ltd., Edinburgh, 1984, 237.
83. **McBride, M. B. and Blasiak, J. J.**, Zinc and copper solubility as a function of pH in an acid soil, *J. Soil Sci. Soc. Am.*, 43, 866, 1979.
84. **Gerritse, R. G. and Van Driel, W.**, The relationship between adsorption of trace metals, organic matter and pH in temperate soils, *J. Environ. Qual.*, 13, 197, 1984.
85. **Schnitzer, M. and Skinner, S. I. M.**, Organo-metallic interactions in soils. V. Stability constants of Cu^{2+}, Fe^{2+}, and Zn^{2+} fulvic acid complexes, *Soil Sci.*, 102, 361, 1966.
86. **Schnitzer, M. and Skinner, S. I. M.**, Organo-metallic interactions in soils. VII. Stability constants of Pb^{2+}, Ni^{2+}, Mn^{2+}, Co^{2+}, and Mg^{2+} fulvic acid complexes, *Soil Sci.*, 103, 247, 1967.
87. **Kerndorff, H. and Schnitzer, M.**, Sorption of metals on humic acid, *Geochim. Cosmochim. Acta.*, 44, 1701, 1980.
88. **Oake, R. J., Booker, C. S., and Davis, R. D.**, Fractionation of heavy metals in sewage sludge, *Water Sci. Technol.*, 17, 587, 1984.
89. Department of the Environment and National Water Council, Report of the Sub-Committee on the Disposal of Sewage Sludge to Land, Standing Technical Committee Report No. 20, National Water Council, London, 1981.
90. **Webber, M. D., Kloke, A., and Tjell, J. C.**, A review of current sludge use guidelines for the control of heavy metal contamination in soils, in *Processing and Use of Sewage Sludge*, L'Hermite, P. and Ott, H., Eds., D. Reidel, Dordrecht, the Netherlands, 1984, 371.
91. **Lake, D. L., Kirk, P. W. W., and Lester, J. N.**, The effects of anaerobic digestion on heavy metal distribution in sewage sludge, *Water Pollut. Control*, 84, 549, 1985.
92. **Steinhilber, P. and Boswell, F. C.**, Fractionation and characterisation of two aerobic sewage sludges, *J. Environ. Qual.*, 12, 529, 1983.
93. **Legret, M., Demare, D., and Marchandise, P.**, Speciation of heavy metals in sewage sludges, in *Proc. Int. Conf. Heavy Metals in the Environment*, Vol. 1, CEP Consultants Ltd., Edinburgh, 1983, 350.
94. **Fraser, J. L. and Lum., K. R.**, Availability of elements of environmental importance in incinerated sludge ash, *Environ. Sci. Technol.*, 17, 52, 1983.
95. **Hodgson, J. F.**, Chemistry of the micronutrient elements in soils, *Adv. Agron.*, 15, 119, 1963.
96. **Lindsay, W. L.**, Inorganic phase equilibria of micronutrients in soils, in *Micronutrients in Agriculture*, Mortvedt, J. J., Giordano, P. M., and Lindsay, W. L., Eds., Soil Science Society of America, Madison, Wisc., 1972, chap. 3.
97. **Salomons, W. and Förstner, U.**, Trace metal analysis on polluted sediments. II. Evaluation of environmental impact, *Environ. Technol. Lett.*, 1, 506, 1980.
98. **Bremner, J. M. and Lees, H.**, Studies on soil organic matter. II. The extraction of organic matter from soil by neutral reagents, *J. Agric. Sci.*, 39, 274, 1949.
99. **Evans, L. T.**, The use of chelating reagents of alkaline solutions in soil organic-matter extraction, *J. Soil Sci.*, 10, 110, 1959.
100. **McKeague, J. A.**, Differentiation of forms of extractable iron and aluminium in soils, *Soil Sci. Soc. Am. Proc.*, 35, 33, 1971.
101. **Hayes, M. H. B., Swift, R. S., Wardle, R. E., and Brown, J. K.**, Humic materials from an organic soil: a comparison of extractants and properties of extracts, *Geoderma*, 13, 231, 1975.
102. **McBride, M. B. and Bouldin, D. R.**, Long-term reactions of copper (II) in a contaminated calcareous soil, *Soil Sci. Soc. Am. J.*, 48, 56, 1984.
103. **Mortvedt, J. J. and Osborn, G.**, Micronutrient concentrations in soil solution after ammonium phosphate application, *Soil Sci. Soc. Am. J.*, 41, 1004, 1977.
104. **Asher, L. E. and Bar-Yosef, B.**, Pyrophosphate, EDTA, and DTPA dependent Zn adsorption by montmorillonite, *Soil Sci. Soc. Am. J.*, 46, 271, 1982.
105. **Bar-Yosef, B. and Asher, L. E.**, Reactions of pyrophsophate in soils and its effect on zinc sorption at various pH levels, *Soil Sci.*, 136, 82, 1983.
106. **Watters, J. I. and Aaron, A.**, Spectrophotometric investigation of the complexes formed between copper and pyrophosphate ions in aqueous solution, *J. Am. Chem. Soc.*, 75, 611, 1953.
107. **Emmerich, W. E.**, Chemical Forms of Heavy Metals in Sewage Sludge-Amended Soils as They Relate to Movement Through Soils, Ph.D. dissertation, University of California, Riverside, University Microfilms, Ann Arbor, Mich., 1980.
108. **Bremner, J. M.**, A review of recent work on soil organic matter, *J. Soil Sci.*, 5, 214, 1954.
109. **Chester, R. and Hughes, M. J.**, A chemical technique for the separation of ferro-manganese minerals, carbonate minerals, and adsorbed trace elements for pelagic sediments, *Chem. Geol.*, 2, 249, 1967.
110. **Chao, T. T.**, Selective dissolution of manganese oxides from soils and sediments with acidified hydroxylamine hydrochloride, *Soil Sci. Soc. Am. Proc.*, 36, 764, 1972.

111. **Le Riche, H. H. and Weir, A. H.**, A method of studying trace elements in soil fractions, *J. Soil Sci.*, 14, 225, 1963.
112. **Lavkulich, L. M. and Wiens, J. H.**, Comparison of organic matter destruction by hydrogen peroxide and sodium hypochlorite and its effect on selected mineral constituents, *Soil Sci. Soc. Am. Proc.*, 34, 755, 1970.
113. **Breward, N. and Peachey, D.**, The development of a rapid scheme for the evaluation of the chemical speciation of elements in sediments, *Sci. Total Environ.*, 29, 155, 1983.
114. **Schalscha, E. B., Morales, M., Vergara, I., and Chang, A. C.**, Chemical fractionation of heavy metals in wastewater-affected soils, *J. Water Pollut. Control Fed.*, 54, 175, 1982.
115. **Sterritt, R. M. and Lester, J. N.**, The value of sewage sludge to agriculture and effects of the agricultural use of sludges contaminated with toxic elements: a review, *Sci. Total Environ.*, 16, 55, 1980.
116. **Rudd, T. and Lester, J. N.**, Investigation of a sequential chemical extraction scheme for heavy metal speciation in sewage sludge, 1st Report to Water Research Centre, Medmenham, U.K., 1985.
117. **Greninger, D., Kollonitsch, V., Kline, C. H., Willemsens, L. C., and Cole, J. F.**, *Lead Chemicals,* International Lead and Zinc Research Organization, New York, 1973, 1.
118. **Rapin, F. and Förstner, U.**, Sequential leaching techniques for particulate metal speciation: the selectivity of various extractants, in *Proc. Int. Conf. Heavy Metals in the Environment,* Vol. 2., CEP Consultants Ltd., Edinburgh, 1983, 1074.
119. **Patterson, J. W. and Hao, S.**, Heavy metal interactions in the anaerobic digestion system, in *Proc. 34th Ind. Waste Conf. Purdue University,* May 1979, Ann Arbor Science, Ann Arbor, Mich., 1980, 544.
120. **Sposito, G., Holtzclaw, K. M., and Levesque-Madore, C. S.**, Trace metal complexation by fulvic acid extracted from sewage sludge. I. Determination of stability of constants and linear correlation analysis, *Soil Sci. Soc. Am. J.*, 45, 465, 1981.
121. **Saar, R. A. and Weber, J. H.**, Fulvic acid: modifier of metal-ion chemistry, *Environ. Sci. Technol.*, 16, 510, 1982.
122. **Riffaldi, R., Levi-Minzi, R., Saviozzi, A., and Tropea, M.**, Sorption and release of cadmium by some sewage sludges, *J. Environ. Qual.*, 12, 253, 1983.
123. **Painter, H. A.**, Metabolism and physiology of aerobic bacteria and fungi, in *Ecological Aspects of Used-Water Treatment, Vol. 2, Biological Activities and Treatment Processes,* Curds, C. R. and Hawkes, H. A., Eds., Academic Press, London, 1983, 11.
124. Commission of the European Communities, Proposal for a council directive on the use of sewage sludge in agriculture, *Off. J. Eur. Communities,* C264, 3, 1982.
125. **Mosey, F. E.**, Methane fermentation of organic wastes, *Trib. Cedebeau,* 34, 389, 1981.
126. **Mosey, F. E.**, Assessment of the maximum concentration of heavy metals in crude sewage which will not inhibit the anaerobic digestion of sludge, *Water Pollut. Control,* 75, 10, 1976.
127. **Mosey, F. E., Swanwick, J. D., and Hughes, D. A.**, Factors affecting the availability of heavy metals to inhibit anaerobic digestion, *Water Pollut. Control,* 70, 668, 1971.
128. **Mosey, F. E.**, The toxicity of cadmium to anaerobic digestion: its modification by inorganic anions, *Water Pollut. Control,* 70, 584, 1971.
129. **Tan, K. H., King, L. D., and Morris, H. D.**, Complex reactions of zinc with organic matter extracted from sewage sludge, *Soil Sci. Soc. Am. Proc.*, 35, 748, 1971.
130. **Baham, J., Ball, N. B., and Sposito, G.**, Gel filtration studies of trace metal-fulvic acid solutions extracted from sewage sludges, *J. Environ. Qual.*, 7, 181, 1978.
131. **Boyd, S. A., Sommers, L. E., and Nelson, D. W.**, Infrared spectra of sewage sludge fractions. Evidence of an amide metal binding site, *Soil Sci. Soc. Am. J.*, 43, 893, 1979.
132. **Hayes, T. D. and Theis, T. L.**, The distribution of heavy metals in anaerobic digestion, *J. Water Pollut. Control Fed.*, 50, 61, 1978.
133. **Petruzzelli, G., Lubrano, L., and Guidi, G.**, Heavy metal extractability from soil treated with high rates of sewage sludges and composts, in *Characterization Treatment and Use of Sewage Sludge,* L'Hermite, P and Ott, H., Eds., D. Reidel, Dordrecht, the Netherlands, 1981, 729.
134. **Sposito, G., Lund, L. J., and Chang, A. C.**, Trace metal chemistry in arid-zone field soils amended with sewage sludge. I. Fractionation of Ni, Cu, Zn, Cd and Pb in soil phases, *Soil Sci. Soc. Am. J.*, 46, 260, 1982.
135. **Harrison, R. M., Laxen, D. P. H., and Wilson, S. J.**, Chemical associations of lead, cadmium, copper and zinc in street dusts and roadside soils, *Environ. Sci. Technol.*, 15, 1378, 1981.
136. **Hickey, M. G. and Kittrick, J. A.**, Chemical partitioning of cadmium, copper, nickel and zinc in soils and sediments containing high levels of heavy metals, *J. Environ. Qual.*, 13, 372, 1984.
137. **King, L. D. and Morris, H. D.**, Land disposal of liquid sewage sludge. II. The effect of soil pH, manganese, zinc and growth and chemical composition of rye *(Secale cereale), J. Environ. Qual.*, 425, 1972.
138. **Boswell, F. C.**, Municipal sewage sludge and selected element application to soil: effect on soil and fescue, *J. Environ. Qual.*, 4, 267, 1975.

139. **Chang, A. C., Warneke, J. E., Page, A. L., and Lund, L. J.,** Accumulation of heavy metals in sewage sludge-treated soils, *J. Environ. Qual.,* 13, 87, 1984.
140. **Schalscha, E. B., Morales, M., Ahumada, I., Schirado, T., and Pratt, P. F.,** Fractionation of Zn, Cu, Cr, and Ni in wastewaters, solids, and in soil, *Agrochimica,* 24, 361, 1980.
141. **Chang, A. C., Pope, A. L., Warneke, J. E., and Grgurevic, E.,** Sequential extraction of soil heavy metals following a sludge application, *J. Environ. Qual.,* 13, 33, 1984.
142. **Schaumberg, G. D., Levesque-Madore, C. S., Sposito, G., and Lund, L. H.,** Infrared spectroscopic study of the water-soluble fraction of sewage sludge-soil mixtures during incubation, *J. Environ. Qual.,* 9, 297, 1980.

Chapter 6

PHYSICAL AND ELECTROCHEMICAL SPECIATION

R. M. Sterritt

TABLE OF CONTENTS

I. Introduction 156

II. Theoretical 157

III. Analytical Approach 158
    A. Physicochemical and Electrochemical Techniques 158
        1. Ion-Specific Electrodes 158
        2. Ion Exchange 159
        3. Polarography 161
    B. Physical Separations 162
        1. Gel Filtration Chromatography 162
        2. Membrane Filtration 164

IV. Heavy-Metal Species in Wastewaters 166

V. Environmental Significance of Speciation 170

References 171

I. INTRODUCTION

Many of the chemical extractions considered in the previous chapter were designed to dissolve from matrices such as sewage sludge and agricultural soils, heavy metals which are bound in the insoluble phase. In sewage and sewage effluent samples, however, considerably more interest is generated by metals in the mobile or soluble phases since it is these which have the greatest potential for environmental dispersion. Clearly, a different analytical methodology is appropriate in matrices such as these. The general approach to physicochemical speciation in wastewater matrices has been adapted from the techniques in use for the characterization of metal forms in "natural waters". There appears to be a greater body of literature associated with natural waters[1-4] than with wastewaters. In assessing the applicability of the various techniques available, it is important to consider the nature of the matrix in each case.

Currently a major area of interest is in the characterization of heavy-metal interactions with organic molecules. This is certainly a more challenging analytical problem than the chemistry of inorganic complex formation[4] with species such as CO_3^{2-}, Cl^-, NH_3, or PO_4^{3-}. Reuter and Perdue,[5] in stressing the importance of organometallic complexes, have suggested that this aspect of heavy-metal speciation has received relatively little attention in the past because (1) it was believed that organic matter in natural waters would be present in negligible concentrations compared to other components of the matrix; (2) the few organic complexing sites available would be dominated by major cations, such as Ca^{2+}, Mg^{2+}, Fe^{2+}, and Fe^{3+}; and (3) there is little scope for the inclusion of ill-defined organic moieties in speciation models.

"Natural" waters may generally be classified as containing <20 mg/ℓ of organic matter,[5] although this arbitrary classification can be extended to include waters with total organic carbon (TOC) concentrations as high as 100 mg/ℓ. A superficial comparison of these typical data with analysis of more complex matrices, such as raw sewage, indicates that concentrations of soluble organic matter may be ten times higher in the latter matrix. Moreover, typical concentrations of suspended matter in sewage (~200 mg/ℓ) are generally much higher than in matrices such as river water.[6] These significant differences in matrix composition may have important implications for the approach to heavy-metal speciation in complex matrices.

It is perhaps not surprising that relatively little attention has been paid to heavy-metal speciation in samples such as sewage, sewage effluent, and sludge due to the incidence of various matrix interference effects and some uncertainty about the validity of sensitive and specific analytical techniques applied to these complex samples. Moreover, the reasons for studying heavy-metal speciation in wastewater may also be somewhat different from the rationale behind studies of natural waters. In natural waters, obtaining data immediately amenable to biological interpretation (i.e., toxicity to aquatic organisms) is probably the most common reason for considering speciation.[4] In wastewater treatment processes, however, approaches to speciation have been relatively unconcerned with direct toxic effects per se, but rather with understanding the partition of heavy metals into various forms or fractions which will ultimately determine their dispersion and impact in the environment. The contrasts between heavy-metal speciation in natural waters and wastewaters and the environmental significance of each may ultimately lead to important conceptual differences in the rationale behind speciation in simple and complex matrices.

The elements which have been considered most often in speciation studies are cadmium, copper, lead, and zinc and, to a lesser extent, nickel, mercury, manganese, chromium, and cobalt.[2] Analytical limitations may be a major reason for the exclusion of other elements from many studies. However, the elements primarily of interest in wastewater matrices will be those recognized as important pollutants[7,8] and those specified in water-quality criteria and sewage sludge disposal legislation.[9]

II. THEORETICAL

Of general significance in wastewater treatment is the distinction between free and bound heavy metals. Binding can be with either soluble ligands or with surface ligands on suspended solids. The association of heavy metals with the suspended fraction has frequently been quantified through the application of adsorption isotherms. The Langmuir isotherm has frequently been applied to metal uptake by activated sludge[10-12] and takes the form

$$\frac{[M_B]}{[S]} = \frac{kb[M_F]}{1 + kb} \qquad (1)$$

where $[M_B]$ = concentration of metal adsorbed, $[M_F]$ = concentration of free metal, k = affinity parameter, b = capacity parameter, and [S] = adsorbent (suspended solids) concentration.

The Freundlich isotherm has the form

$$\frac{[M_B]}{[S]} = k[M_F]^{1/n} \qquad (2)$$

where k and n are also empirical constants. This isotherm has been applied only rarely in wastewater matrices.[13]

The association of heavy metals with organic ligands, both soluble and surface bound, may in many cases be via the formation of complexes. The equilibrium relationship between $[M_F]$ and $[M_B]$ for complexation phenomena involving 1:1 stoichiometry is analogous to the Langmuir isotherm in Equation 1, thus permitting adsorption and complexation data to be treated in the same way. The equilibrium reaction for complex formation is

$$aM + bL \rightleftharpoons M_aL_b$$

and the equilibrium constant (K) is given by

$$K = \frac{[M_aL_b]}{[M]^a[L]^b} \qquad (3)$$

where all concentrations are those at equilibrium. Where a = b = 1, the important parameters can be evaluated experimentally in terms of $[M_F]$ and $[M_B]$. The conditional stability constant is then given by

$$K' = \frac{[M_B]}{[M_F]([L_T] - [M_B])} \qquad (4)$$

For a range of values of $[M_F]$ and $[M_B]$ a plot of $[M_F]/[M_B]$ against $[M_F]$ will yield a straight line from which K' and $[L_T]$ can be calculated. The equation of the line is

$$\frac{[M_F]}{[M_B]} = \frac{M_F}{[L_T]} + \frac{1}{K'[L_T]} \qquad (5)$$

This is analogous to the linear form of the Langmuir isotherm which can be derived from Equation 1.

Ruzic[14] has extended the theoretical basis of complexation to include biligand systems, including systems where the two ligands have very similar affinities for the heavy metals, and systems where the complex formed with one ligand is indistinguishable analytically from the free metal. These theoretical models have not been applied to wastewaters, however.

Another relationship which may be applicable in wastewaters especially for binding by macromolecular ligands is the Scatchard Equation:[15]

$$K' = \frac{M_B}{L_T[M_F]} \left(\frac{1}{n - \frac{M_B}{L_T[M_F]}} \right) \quad (6)$$

where M_B is the number of moles of metal bound by L_T, the number of moles of the ligand molecule present, and n is the number of metal binding sites on the molecule.

III. ANALYTICAL APPROACH

The primary approach to studying heavy-metal speciation in any type of sample is to identify the species of interest[14] and to choose analytical techniques which are selective for these species. There are then two further alternative approaches involved in this specific determination. The first is to use an analytical technique which is highly specific for one species or one group of species in the presence of many other metal forms and to make a direct determination on the sample. The second alternative is to employ a separation procedure to isolate the species of interest prior to determining the metal concentration, possibly by a less specific analytical method, in the separated component.

A. Physicochemical and Electrochemical Techniques
1. Ion-Specific Electrodes

Direct determination by ion-specific electrodes (ISEs) has been used by several workers for the determination of free metal (M^{n+}) in systems where a variety of other metal species exist. From a practical point of view, ISEs are convenient tools for speciation since calibration is simple and an almost direct readout of the free metal ion activity (which is almost equal to its concentration at low concentrations) is obtainable after a short period of equilibration. However, the observations of some workers studying both simple and complex matrices have raised doubts about the selectivity of the electrodes for the metal ions of interest. It is well known that some electrodes which are ostensibly specific for one metal will respond to other metals present in excess of certain limiting concentrations.[16] Sometimes, potential problems of this nature may be overcome by using a general method of analysis (e.g., atomic absorption spectrophotometry) to ensure that interfering ions are not present at significant concentrations. In sewage and sewage effluent samples, most of the metals will be present within the same concentration ranges, and so likely interferences are, to a degree, predictable.

Interferences of an unknown nature in the determination by ISE of [Cu^{2+}] in hard lake waters have been noted by Barica.[17] The slopes obtained when the samples were titrated with Cu^{2+} were approximately twice the value expected. This would theoretically occur if the titrant behaved as a monovalent cation, since the electrode potential, E, is given by

$$E = E_o + \frac{2.3RT}{nF} \log a \quad (7)$$

where R, T, and F are constants, a is the activity ($\simeq [M^{n+}]$), and n the charge on the ion. Barica[17] suggested that the response may have been elicited by monovalent complexes of copper of the type $CuOH^+$ and $CuHCO_3^+$ which could be present in sewage matrices in hard-water areas and could lead to erroneously high values for the determination of $[Cu^{2+}]$.

ISEs may respond to other species, unrelated to the metals of interest. El-Taras and Pungor[18] found that a Cu^{2+}-selective electrode gave a positive response in solutions containing citrate, tartrate, glycine, or serine in the absence of copper and suggested that this was due to the potential generated by the complexation of the electrode membrane itself (often incorporating a crystalline lattice of copper selenide or copper telluride). The stronger ligands elicited a response at concentrations down to 10^{-6} M, while the weaker ligands had an effect at about 10^{-2} M. The stability constants of metal complexes of these ligands[19] may be similar to those of wastewater ligands. The use of ISEs in the presence of undefined organic ligands having similar complexing characteristics could therefore result in significant errors in the analysis. Where pH buffers are required, these should also be checked for possible interferences.[20]

In studies of cadmium speciation in a bacterial culture medium,[21] erroneously high responses of a Cd^{2+}-specific electrode were found.[22] Sodium citrate was used initially at a concentration of 7×10^{-3} M as the sole carbon source for bacterial growth. In the presence of citrate, but with no added Cd^{2+}, the electrode showed a positive response, even when the citrate concentration was reduced to 10^{-6} M. In a solution containing 5×10^{-7} M Cd and 10^{-6} M citrate, 88% of the metal was detected as Cd^{2+}, while raising the citrate concentration to 2×10^{-6} M led to 94% of the metal apparently existing in the free ionic form at pH 6.8. Gel filtration chromatographic analysis[21] and the application of a computer model[23] demonstrated that the metal was probably almost completely complexed by citrate under these experimental conditions.

One potential advantage of ISEs is that, unlike many other methods, they may be used directly on samples containing suspended solids, which otherwise may have to be removed from the sample, with unknown consequences for changing speciation equilibria. van den Meent et al.[24] used a Cd^{2+}-selective electrode to monitor the titration of suspended solids which had been recovered and resuspended from the original sample. Under these conditions, it may be expected that surface-bound ligands would produce less interference than soluble complexing agents. Determination of the stability constants of copper and lead complexes with suspended activated sludge biomass using ISEs was found to be reasonably straightforward,[20] although the analysis was undertaken in the absence of any potentially interfering components of the soluble phase which had been separated by filtration prior to resuspension of the suspended solids. Jardim and Allen[25] examined copper complexation in secondary sewage effluent using an ISE and found that the results compared well with those obtained by two other methods for the calculation of K' and $[L_T]$. There was no evidence of interference in the analysis.

Gardiner[26] used a Cd^{2+}-specific electrode to determine the free ion in settled sewage, sewage effluent, and other aqueous samples and reported no interferences, despite the predicted existence of between 5 and 15% of the metal in the forms $CdOH^+$ and $CdCl^+$ and the presence of soluble organic ligands. It is possible that under certain circumstances, electrode interferences would go undetected; however, it was found that experimentally determined values of $[Cd^{2+}]$ in a range of samples agreed well with predicted values based on thermodynamic data.[26]

2. Ion Exchange

A method for determining $[Ni^{2+}]$ in raw sewage by employing its selective adsorption by a strong acid cation exchange resin (Dowex® 50W-X8) has been described by Cantwell et al.[27] The determination was undertaken in batch, by equilibrating the sample with the resin at pH 8.2 for 4 hr, or by passing the sample slowly through a column of the resin. The method depends on the assumption that $[Ni^{2+}]$ is the only species which is adsorbed onto the resin. Free metal ion concentrations can then be determined by

$$[Ni^{2+}] = \frac{[R_2Ni]}{\lambda_o} \qquad (8)$$

where $\lambda_o = [R_2Ni]/[Ni^{2+}]$ in the absence of the ligand (i.e., when a pure solution of Ni[NO$_3$]$_2$ is passed through the column) and is a constant under the experimental conditions used. The batch method used by Cantwell et al.[27] required that the sample be well buffered with respect to free metal ion (i.e., $[L_T] \gg [M_T]$) so that adsorption of the small amount of $[Ni^{2+}]$ present would not significantly alter its concentration in the sample. The column method, however, does not require this condition since the sample may be passed through until no further Ni^{2+} is adsorbed, such that the sample is in true equilibration with the resin. This latter method, which was generally preferred, allowed the determination of $[Ni^{2+}]$ from Equation 8 by desorbing and determining the quantity of the metal adsorbed onto the resin.

Using a batch-type of equilibration with ion-exchange resin, the total metal remaining in solution may be determined. This includes the fraction of M^{n+} not adsorbed by the resin, plus the complexed metal (ML_x). Thus, the distribution coefficients for a given metal in the absence and presence of ligands are

$$\lambda_o = \frac{[R_nM]}{[M^{n+}]} \qquad (9)$$

and

$$\lambda = \frac{[R_nM]}{[M^{n+}] + [ML_x]} \qquad (10)$$

From these values conditional stability constants of the complex formed may be obtained from the relationship

$$\log\left(\frac{\lambda_o}{\lambda} - 1\right) = \log K + x\log[L] \qquad (11)$$

This linear function can be solved graphically for x and K.[28] The ion-exchange method of Cantwell et al.[27] gives a value for [L] in the form of the complexing capacity of the sample. From values for $[M^{n+}]$ generated from Equation 7, a plot of $[M^{n+}]$ vs. $[M_T]$ yields a biphasic curve, with the lower part having a small gradient and the upper part, corresponding to metal added after saturation of the available binding sites, having a gradient of 1.

The assumption that Ni^{2+} is that only species adsorbed during ion exchange[27] was justified from the observations that (1) principal organic ligands in sewage are likely to be multidentate or carry multiple negative charges, and that neutral ligands, such as NH_3 are not likely to be present in large concentrations; and (2) that the exchange resin has a lower affinity for species with a lesser charge than Ni^{2+} (e.g., NiL^+). Laxen and Harrison[29] however, used Chelex-100 in the calcium form and preferred the term "Chelex-labile" metal for the sum total of species adsorbed during batch equilibration for 48 hr. The contention that ion-exchange resins may have a lower affinity for ML^+ than for M^{2+} is in a sense subjective, since whether or not a complex is adsorbed depends on the relative strengths and, to a degree, the relative concentrations of the ion exchanger-metal complexes and the ligand-metal complexes.

One important characteristic of wastewater samples is the probable existence of a large number of ligands, which may have similar complexation characteristics. In this case the stability constant measured will be a composite function of the stability constants of the many individual complexes formed. Using the ion-exchange method for measuring the apparent stability constant, K_S, in a system containing two ligands, L_1 and L_2, Crosser and Allen[30] have shown that since

$$\frac{\lambda_o}{\lambda} - 1 = \frac{[ML_1] + [ML_2]}{[M^{n+}]} \quad (12)$$

then

$$\log\left(\frac{\lambda_o}{\lambda} - 1\right) = \log([L_1] + [L_2]) + \log\frac{K_1[L_1] + K_2[L_2]}{[L_1] + [L_2]} \quad (13)$$

such that

$$K_S = \frac{[L_1]K_1}{[L_T]} + \frac{[L_2]K_2}{[L_T]} \quad (14)$$

This relationship can be extended to cover multiligand systems, ML_1, ML_2, ML_3, ... ML_N, having stability constants K_1, K_2, K_3, ... K_N, where

$$K_S = \sum \frac{[L]K}{[L_T]} \quad (15)$$

This relationship is potentially useful in describing heavy-metal complexation in complex samples, especially if the characteristics of one of the ligands present may be studied separately from the others.

3. Polarography

Many methods applied to complex matrices will not detect only free metal ions. Polarographic methods will generally detect free metal ions, which are reducible at the mercury electrode, together with certain metals present as "labile" complexes. "Labile" complexes are considered to be those which are rapidly dissociable at the mercury electrode under the experimental conditions used. The quantity of metal deposited at the electrode will depend on the rate of dissociation of its complex.[2] This introduces a further variable, the *rate* of dissociation of a complex, in addition to its stability constant and its complexation capacity. Whether or not this rate constant (k_{-1}) has any functional relationship with the inherent strengths and distributions of metal complexes in a given sample is at present uncertain.

Laxen and Harrison[31] have applied anodic stripping voltammetry (ASV) to the speciation of heavy metals in sewage effluent, expressing the results as labile or nonlabile metal. Bender et al.[32] also applied ASV to sewage effluent which had been fractionated by gel filtration chromatography in order to establish whether "free" metal was present in any of the separated fractions.

Because of the operational limitations associated with the concept of polarographically labile complexes, any data which are obtained regarding complexation characteristics of complex matrices will not be analogous to data obtained by other methods. However, the stability constants and complexation capacity of nonlabile complexes may be determined simply by titration with free metal ions and measuring the concentration of labile metal.[33]

The inability of polarography to differentiate between labile complexes and free metal could have serious implications for the accurate speciation of heavy metals. O'Shea and Mancy[34] have described how the results of titrimetric analysis using ASV could distinguish between nonlabile complexes, labile complexes, and free metal. From titrations of a solution of the metal ion with the ligand, the constancy of peak current measurements with shifts in peak potential values indicates labile complex formation; whereas reductions in peak current with successive additions of the ligand are indicative of nonlabile complexes. Using the metal ion as the titrant, instead of the ligand, will serve to differentiate between nonlabile complex formation with the supporting electrolyte and labile complex formation with the ligand itself. Sposito[35] has suggested overcoming this problem by using ClO_4^- as a relatively inert supporting electrolyte.

In wastewater matrices, where both labile and nonlabile complexes are likely to be present in the same sample, differentiation of the species in this manner may prove to be impossible, and the simple distinction between total (M_T), nonlabile (equivalent to a portion of M_B) and labile (equivalent to a portion of M_B plus M^{n+}) from a direct measurement may have to suffice. Moreover, the interferences which may occur in largely uncharacterized samples have been well documented. The possibility of irreversible reduction of nonlabile complexes is a serious drawback. In considerations of how the lability of a complex relates to its stability has sometimes been the implication that nonlabile complexes are thermodynamically stable. However, under certain conditions Cu-nitrilotriacetate and Cu-EDTA complexes have been found to be reducible.[36] Other types of interference commonly recorded are particularly prone to occur in wastewater samples where high concentrations of organics are present since these can adsorb directly onto the electrode, producing erroneous results.

Interferences in the differential pulse ASV (DPASV) determination of copper in the presence of fulvic acid,[37,38] originally suspected to be due to adsorption, were more probably due to complex lability. A low value obtained for [L_T] in the DPASV analysis of copper complexation in secondary sewage effluent[25] could have arisen from lability. Varney et al.[39] have shown that lead complexes with naturally occurring organic matter are ASV labile, whereas in a sample containing more than one type of binding site, only some of the copper complexes may be labile.

B. Physical Separations

Physical separation of heavy-metal species is an attractive approach for complex matrices.[2] Wastewater samples are more "physically heterogeneous" than, for example, natural waters, and physical separation of the components may afford considerable insight into the distribution of metal species. Physical separation is, effectively, differentiation on the basis of size. Again, widespread application of these techniques has not been made.[2,3] However, the methodology has been considered in a comprehensive review,[40] although again the emphasis was placed on natural waters.

1. Gel Filtration Chromatography

Macromolecular complexes of heavy metals in the soluble phase may be separated and characterized to some extent by gel filtration chromatography. The use of gel filtration chromatography for size separation of metal species in the soluble phase was first applied to sewage effluent by Bender et al.[32] and to sludge-soil mixtures by Mercer and Richmond.[41,42] Since this early work, the technique has not been favored for the study of complex matrices, although it has found application for some natural waters[43] and extracts of sewage sludge.[44] However, gel filtration has been employed recently in further studies of cadmium in bacterial culture media,[21] mechanisms of metal removal in laboratory-scale activated sludge simulations,[45,46] and metal forms in secondary sewage effluent.[47] Heavy-metal speciation by gel filtration has been discussed at some length in two recent reviews.[4,40] One major drawback of this procedure is the significant dilution of the sample during its passage through the column which led Florence and Batley[48] to discount it as a viable method for the majority of natural waters. In general, separation efficiency deteriorates at sample loading volumes

(V_L) of greater than 3% of the column volume (V_T). Since typically the total volume of eluent collected from chromatographs of complex samples is three times the column volume, in the hypothetical case of an equal distribution of the metal between all of the eluted fractions, the mean dilution of the determinand will be about 1:100. This represents an extreme case, since, in many water samples, the metal only appears in a few eluted fractions.[40] However, in a sewage effluent sample chromatographed on Sephadex G-50, some copper appeared in about 70% of all eluted fractions,[47] so that a dilution of 1:50 to 1:75 should be assumed. The metal concentration in the sample to be applied should therefore be at least 50 times the detection limit of the analytical technique used to determine it in the eluted fractions. This condition can often be achieved in the case of the more soluble metals at typical concentrations in sewage and sewage effluent samples. Where it is not, preconcentration of the sample prior to its application to the column should be avoided, as in all speciation techniques, since this tends to exaggerate the importance of organic complexes.[49]

One problem associated with gel filtration is that charged species sometimes behave anomalously, either being retarded to some extent or eluted rapidly. This can occur because of the weak ion-exchange characteristics of Sephadex, one of the most commonly used gels. During calibration of a gel column, the use of a soluble metal salt (typically $M(NO_3)_n$) to determine the salt volume often gives values of $V_s \gg V_T$, which is clearly untenable purely on the basis of size separation. Typical values of V_s for cadmium, copper, nickel, and manganese on Sephadex columns exceed V_T by factors of 2 to 3.[45,47] However, this adsorption appears to be completely reversible, since approximately 100% recovery of most cations occurs after equilibration of the gel with the sample or by passage of metal ion solutions, unless decontamination of the gel with nitric acid is attempted, where this appears to activate stronger binding sites resulting in up to 70% losses of ionic metal spikes applied to the column.[45] The adsorption characteristics of gel media can be suppressed to some extent by using an eluent with an ionic strength of 0.02 to 0.05 M, usually provided by alkali metal chlorides or nitrates. In view of the effects of divalent cations on reducing adsorption of heavy metals during membrane filtration,[29] it may be possible to obtain a greater suppression of the selective adsorption of divalent heavy metals by using calcium or magnesium salts. Posner,[50] in studying adsorption effects on Sephadex, found that G-25 was the grade of gel which exhibited the least adsorption effects.

Low-molecular-size species of heavy metals which are appreciably anionic are likely to be excluded during gel filtration such that they give the appearance of having high molecular weights. $Cr_2O_7^{2-}$ is an example of such a species which may be eluted close to the void volume.[51] However, high-molecular-weight complexes having only a slight net charge (ML^+ or ML^-) may be less prone to charge interactions with weak ligands located in those phases of the gel which are accessible only to small molecules. Generally, if gel filtration is confined to metals which form predominantly cationic inorganic species at slightly acid pH, it may be surmised that metal peaks eluted slightly ahead of $V_{s(M^{n+})}$ are not of the type M^{n+} and species eluted significantly ahead of $V_{s(M^{n+})}$ are probably associated with macromolecular organics.

Further confirmation of the association of metal fractions partially or totally excluded from the gel may be obtained from measuring the ultraviolet absorbance or organic carbon (as TOC or chemical oxygen demand [COD], for example) of the fractions in order to demonstrate the coincidence of macromolecular organics and metals. Baham et al.[44] were able to demonstrate the existence of both inorganic and organic forms of metals in samples of fulvic acid isolated from sewage sludge/soil mixtures by correlation of the metal fractions eluted from Sephadex G10 with TOC concentrations. However, these authors pointed out that the precise coincidence of metals and organics may not occur due to displacement phenomena caused by incomplete equilibrium or slight dissociation of weaker complexes. In view of the observation of Means et al.[49] regarding the effects of sample concentration, it may be expected that the dilution of a sample passing through a column would have an effect on some complex equilibria.

Although the precise effects of gel chromatography on disturbing equilibrium of the sample are unknown, any potential problems have been circumvented by Mantoura and Riley[15] in establishing the equilibrium required for the determination of stability constants of metal-fulvic acid complexes. These authors added a range of concentrations of copper to solutions of fulvic acid and eluted the sample from a column of Sephadex G-15 using an eluent containing the same concentration of Cu^{2+} as was added to the sample. Thus the complex was exposed to a constant free metal ion concentration during its passage through the gel column, and from a plot of total copper against elution volume, the return of $[Cu_T]$ to baseline values of $[Cu^{2+}]$ between the initial peak (equivalent in area to M_B) and the subsequent trough (also equivalent in area to M_B) was indicative of the attainment of equilibrium. The results could be treated according to a Scatchard equation.

The experiments of Mantoura and Riley[15] were conducted using a reasonably homogeneous sample whose elution characteristics were quite different from those of the free metal ion. If a complex sample were to be analyzed directly, it may be possible to simultaneously evaluate the complexation parameters of several separated peaks, but the free metal ion trough at V_s may be obscured or difficult to evaluate. Provided that equilibrium conditions are not affected significantly by eluting samples directly, without exposure to constant concentrations of $[M^{n+}]$, it may be possible to obtain complexation parameters from a series of direct separations, although again the determination of $[M^{n+}]$ it likely to be difficult. Since gel filtration, in common with other speciation methods such as ASV, will not reliably distinguish $[M^{n+}]$ from certain other complexed species, there may be a use for a method of treating titration results which takes into account the nonspecific nature of some analytical techniques for $[M^{n+}]$.

2. Membrane Filtration

Although the approach towards speciation of heavy metals in the soluble phase is well established, in samples such as sewage and sewage sludge, insoluble forms of heavy metals can dominate the distribution of species. Even in sewage effluent, between 10 and 62% of the major heavy metals may be in insoluble forms.[52] Insoluble forms have been studied to a much lesser extent than the soluble phase, principally because they are of lesser toxicological significance and they are not as amenable to study. However, the range of types of particulate matter which heavy metals may become associated with in wastewater matrices is probably as complex as the soluble phase ligands.

The obvious method of separating particulates is by filtration. A single filtration through a 0.45-μm membrane filter will arbitrarily separate "soluble" from "insoluble" species based on a commonly accepted size definition,[40] although smaller filters have been used for wastewater matrices in order to exclude bacterial solids completely.[53] A broader size separation may be obtained using a range of filter pore sizes. Laxen and Harrison[29] used five different filters in the range 0.015 to 12 μm, in studies of sewage effluent, while Kempton et al.[54] used six filter sizes in the range 0.2 to 25 μm in studies of activated sludge and sewage effluent. Although such an approach will provide a more comprehensive speciation of particulate forms of heavy metals, the actual differentiation between physical types will probably be quite difficult. Bacterial cells can be in the size range of >0.2 μm, inorganic precipitates may exist as particles in the range 0.3 >1 μm,[55] while larger colloids can approach the lower end of this size range. Biological solids in the activated sludge process tend to exceed 20 μm in size.[54] Moreover, there may be the additional complication that association of heavy metals with such particles is not simply a metal-surface ligand interaction and a number of more complicated interactions may occur.[56]

From a practical point of view, the most common drawbacks of filtration will be adsorption of soluble metal species onto the filter (or filter holder) and clogging of the pores under conditions of high solids loading. The first problem may be overcome by dilute acid washing techniques, saturation of binding sites with divalent cations, and equilibration of the filter with fairly large volumes of sample prior to collection of the filtrate.[40] The latter precaution may prove to be impractical, since in samples with high suspended solids concentrations, equilibration with the sample may only serve to clog the filter before filtrate collection may take place. Circumvention of the problem of clogging may be achieved by centrifugation of the sample prior to filtration of the supernatant,[57] although this may introduce errors where a range of filters are used for size fractionation. Although filter clogging may be reduced by stirring the sample during filtration, preventing the formation of a surface deposit, Laxen and Chandler[58] reported that no particular benefit was obtained from using stirred cells. Prefiltration through large-pore-size filters may be effective in removing larger particles which could block microporous filters, but the use of a series of filtration steps may increase the opportunity for contamination.[54]

Although the adsorption of metals onto solids has been described in similar fashion to the formation of soluble ligand complexes[24] by a form of Equation 5, generally such adsorption is described by different concepts. Cheng et al.[28] and Patterson and Hao[59] determined metal-sludge solids stability constants, given by

$$M + yS \rightleftharpoons MS_y; \quad K = \frac{[MS_y]}{[M][S]^y} \tag{16}$$

from measurements of the solids-bound metal after separation of the soluble phase of activated sludge mixed liquor. The following relationship was derived

$$\log\left(\frac{[M_B]}{[M_F] + [ML]}\right) = \log K + y\log[S] \tag{17}$$

where [ML] corresponds to the soluble phase complexes described by Equation 10.

Gould and Genetelli[60] used a competitive form of the Langmuir isotherm, which included a parameter for pH to account for protonation of the ligands, in a study of metal binding by anaerobically digested sewage sludge. This form was given by

$$\frac{1}{[M_B]} = \frac{1}{[L]} + \left(\frac{1}{b[L][M_F]}\right)\left(1 + \frac{[H^+]}{D}\right) \tag{18}$$

where D is a constant. Values for b, D, and [L] by were obtained by regression analysis of the linear form of the isotherm fitted to experimental data for $[M_B]$ and $[M_F]$. The sludge, equilibrated with the added heavy-metal solution, was centrifuged and the supernatant filtered through an ultrafilter having an effective pore size of 2.0 nm, which effectively excluded colloidal materials. Even so, the authors were careful to point out, the measured values of $[M_F]$ would include metal associated with numerous low-molecular-weight ligands. Stability constants (K_o) for the metal-sludge complexes were obtained from the relationship

$$\frac{1}{[M_B]} = \frac{1}{[L]} + \frac{1}{K_o[L][M_F]} \tag{19}$$

where

$$K_o = \frac{bD}{D + [H^+]} \tag{20}$$

The problems which may arise in filtration or ultrafiltration due to contamination, losses by adsorption, and poor selectivity for particles of a given size have all been discussed at length with reference to natural waters.[40] In certain cases, these factors may be more critical in wastewater matrices than in natural waters, due to the generally higher concentrations of organic matter and suspended solids. Polycarbonate filters, recommended for natural water filtration, may be unsuitable for use in organic-rich matrices because they adsorb humic substances.[61] Particularly severe problems with clogging of filters with suspended solids may occur;[57] problems encountered with natural waters[62] would suggest membrane filtration of many wastewater matrices to be impractical.

In common with certain other speciation methods, complex equilibria may be affected by transient changes in the concentration of metal species during filtration. Overestimates of the complexation capacity of the solid phase may occur due to the increase in solids concentration above the filter relative to $[M_F]$ as the filtration proceeds.[20]

IV. HEAVY-METAL SPECIES IN WASTEWATERS

Although many of the methods of heavy-metal speciation described above have met with only limited success when applied to complex matrices, the limited available data suggest that speciation may be very much different in wastewater samples than in "natural waters". This may be particularly true where the species M^{n+} is concerned.

Cantwell et al.[27] developed their ion-exchange technique specifically for the determination of $[Ni^{2+}]$ in sewage but, in six out of seven samples studied, could find none at concentrations greater than $5 \times 10^{-9}\,M$ (i.e., about 0.3 µg/ℓ). In these samples $[Ni^{2+}]$ represented about 1% of the total "soluble" nickel, which was determined by centrifugation. Bender et al.[32] also found no "free" copper in a sample of sewage effluent, as determined by ASV. The absence of free copper was apparently inferred from the absence of a stripping peak which would probably preclude the presence of Cu^{2+} at concentrations above the detection limit of ASV. Patterson[63] in seven replicated analyses of a sewage effluent found that the free Cu^{2+} concentration varied between 0.8 and 3.2% of the total soluble copper. If insoluble species are taken into account, the proportion of the total copper in a sample existing in the form Cu^{2+} would be very small indeed. In the case of nickel and copper, therefore, speciation methods which place an emphasis on the M^{n+} form would be attaching an importance to this species out of proportion to its relative abundance. This inability to find M^{n+} is indicative of a system with a large complexation capacity $[L_T]$ and containing relatively stable complexes. Complexation capacities of sewage[27,64] and sewage effluent samples[25,31,32,63,65,66] reported in the literature are given in Table 1. Also included in Table 1 are typical values of $[M_T]$. These data indicate that for those heavy metals where the soluble phase complexation capacity of the sample exceeds $[M_T]$, significant removal in biological wastewater treatment will depend on the dissociation of these complexes to form settleable species. It would appear that copper and nickel would in many cases be largely present as complexed forms in sewage.

Values of cadmium and lead complexation capacities given in Table 1 were determined largely from ASV,[31,66] and their low values compared with values for copper and nickel may be related to the techniques used. However, using a Cd^{2+}-specific electrode, it was found that 35 to 41% of the cadmium in three wastewater samples was in the free ionic form,[26] perhaps suggesting that removal of this metal would be largely dependent on the direct association of Cd^{2+} with the activated sludge biomass. Laxen and Harrison[31] were unable to quantify labile responses for lead and cadmium in sewage effluent but found that 58% of the cadmium was ASV "labile", indicating that ASV may be limited in its application to the study of complexation in certain matrices. A similar proportion (62%) of the cadmium was also found to be Chelex "labile", again indicating that perhaps this fraction included

Table 1
HEAVY-METAL COMPLEXATION CAPACITIES OF SEWAGE AND SEWAGE EFFLUENT

Sample	Metal	Complexation capacity (μM)	Method	Typical $[M_T]$ (μM)	Ref.
Raw sewage	Cu	53	Cu(OH)$_2$ solubilization	3	64
	Ni	10	Ion exchange	4	27
Settled sewage	Cu	47	Cu(OH)$_2$ solubilization	1.5	64
Sewage effluent	Co	4.1	ASV	0.5	65
	Cu	1.0	ASV	1.0	66
	Cu	2.0	ASV	1.0	31
	Cu	9.0	Equation 27	1.0	63
	Cu	1.0—80	ASV	1.0	32
	Cu	17.2	ISE	1.0	25
	Cu	3.60	MnO$_2$	1.0	25
	Cu	2.05	ASV	1.0	25
	Pb	0.077	ASV	0.02	66
	Pb	0.02	ASV	0.02	31
	Cd	ND[a]	ASV	0.05	66
	Cd	ND	ASV	0.05	31

Note: ASV, anodic stripping voltammetry; ISE, ion-specific electrodes.

[a] ND, not detectable specified only as "sewage samples".

significant quantities of complexed species, and that the stabilities of complexes when exposed to these analytical techniques would not be an accurate reflection of their physical stabilities in the sample matrix itself.

Conditional stability constants for copper and nickel binding by soluble ligands in the mixed liquor from an activated sludge plant have been obtained using an ion-exchange technique.[28] However, in using Equation 10 for data interpretation, true values of K' could not be obtained since an accurate estimate of $[L_T]$ was not available, with the COD of the soluble phase being substituted instead. Further work,[63] conducted in a similar manner, also produced conditional constants based on the substitution of TOC for the true value of $[L_T]$. The data from these two studies[28,63] are, however, significant for comparative purposes. Conditional stability constants for copper and nickel complexation by the soluble phase ligands were of the same order of magnitude, whereas the corresponding values of K' for binding to the solids differed by almost three orders of magnitude, with nickel binding being the weaker of the two. This phenomenon may help to explain the widely differing behavior of the two metals in biological wastewater treatment.

The binding capacity of activated sludge solids is largely attributable to bacterial extracellular polymers.[67-69] Conditional stability constants of complexes between several heavy metals and extracted extracellular polymers of activated sludges[13,70,71] and the sludge itself[11,62,70] have been determined by a variety of physical separation techniques. A summary of the values of K' obtained is shown in Table 2. In many cases, more than one major group of binding sites have been identified. For comparative purposes, however, only values of $[L_T]$ and K' are included, these being calculated as total complexing capacity and as a weighted average (K'_s), respectively. K'_s values representing this weighted average were calculated from Equation 14. The formation of stronger complexes by copper than by nickel is again a fairly consistent finding, with cobalt complexes generally being the weakest. Complexation capacities of activated sludge solids are, however, similar for the different metals, corre-

Table 2
SUMMARY OF STABILITY CONSTANTS AND COMPLEXATION CAPACITIES REPORTED FOR METAL BINDING TO ACTIVATED SLUDGE SOLIDS AND EXTRACTED POLYMERIC FRACTION

Metal	$\log_{10}K'$ Sludge	$\log_{10}K'$ Polymer	$[L_T]$, μmol/g Sludge	$[L_T]$, μmol/g Polymer	Ref.
Cd	5.9	4.9	86	210	12[a]
	—	7.5	—	290	13[a]
Co	5.2	4.7	69	57	12[a]
	—	5.9	—	91	13[a]
Cu	6.7	5.8	100	100	12[a]
	—	7.5	—	290	13[a]
	6.2	—	19	—	11
	6.8	—	25	—	20[a]
Mn	5.9	—	10	—	11
Ni	5.2	4.9	83	29	12[a]
	—	6.3	—	550	13[a]
	5.8	—	3	—	11
Pb	6.0	—	25	—	20[a]

[a] Values of K' reported are K'_s values calculated using Equation 14, representing overall metal-binding properties of the system.

sponding to approximately 1 to 20 mg/ℓ in a typical mixed liquor, which although sufficient in most cases to account for most of the metals present at typical concentrations, do not reflect the binding potential of extracellular polymers.

Differences in the affinities of metals for activated sludge solids are not reflected in their binding to primary and digested sludge solids. Values of K' found for copper and nickel binding were of the same magnitude in primary sludge[63] and nearly identical in anaerobically digested sludge,[60] although an order of magnitude below K' for activated sludge. Cadmium and zinc binding in anaerobically digested sludge also had similar stabilities to the other metals studied.[60] In the soluble phase of anaerobically digested sludge, however, very large differences in the stabilities of soluble complexes were observed.[63]

Although it is probable that very small proportions of heavy metals in the soluble phase of wastewater matrices are in the form M^{n+}, the precise nature of the complexed forms are unknown, but it appears that they are largely organometallic species. Patterson[63] suggested that inorganic complexes account for only 10 to 15% of the copper in sewage effluent, with most of the remainder being associated with organic carbon. Rossin et al.,[47] using a gel filtration technique, found significant quantities of copper and nickel associated with fractions representing molecules with a molecular weight of 10^4 to 3×10^4 in samples of settled sewage. The remainder of the metals appeared in fractions corresponding to a molecular weight of $<10^3$ which may have included organic complexes, inorganic complexes, and free forms of the metals. Generally, this lower-molecular-weight fraction was absent in the final effluent, whereas the high-molecular-weight fraction persisted throughout the treatment process. A greater proportion of nickel than of cadmium was associated with the lower-molecular-weight fractions, suggesting a degree of specificity for certain ligands. Sterritt and Lester,[45] studying an activated sludge system maintained on synthetic sewage, found by gel filtration that both copper and manganese were associated with high-molecular-weight fractions but that the two metals appeared to be complexed by different and discrete macromolecular components of the synthetic sewage. The different speciation patterns observed appeared to be reflected by the significantly different removal efficiencies of these two metals.[72]

In a subsequent study,[46] using Sephadex G-25 to fractionate sewage effluent from activated sludge simulations, fairly complex elution profiles were obtained. In general, copper and manganese were associated with fractions corresponding to molecular weights of <5000. Cadmium, cobalt, nickel, and thallium were found to be increasingly associated with higher-molecular-weight fractions at longer sludge ages, consistent with the observations that macromolecular slime material is released from the floc matrix at slow growth rates[73] and that this can have significant metal-binding properties.

Additional evidence of ligand specificity was obtained from studies on functional group blocking of anerobically digested sludge solids by methylation.[74] The metal-binding capacity of the sludges was determined by saturating the dried solids in a column followed by washing with distilled water. Methylation of one sludge sample caused reductions in zinc, copper, nickel, and cadmium binding capacities of 38, 48, 8, and 16%, respectively. In another sample, methylation had virtually no effect on zinc and nickel binding, but caused reduction of 22 and 44% in the binding of copper and cadmium, respectively. This suggests that nickel is not predominantly associated with the carboxylic or phenolic functional groups which are the targets of methylation.

Infrared analysis of air-dried sewage sludges,[74] in fact, revealed that the concentration of carboxylic functional groups was minimal, but that there was evidence from discrete shifts (in response to metal addition or removal) in the wave number of the peak associated with amide groups that this was involved in heavy-metal binding.

The accurate speciation of heavy metals in wastewater matrices is complicated by the fact that the idealized system of metal-ligand interactions is difficult to apply when precipitation of the metals occurs. Ruzic,[14] in describing methods whereby soluble metal complexation may be studied, used the assumption that heavy metals in the form of precipitates would not dynamically participate in metal-ligand systems; this assumption is convenient for soluble phase speciation modeling. However, in complete contrast, the use of a copper hydroxide precipitate ($Cu[OH]_2$) to measure the complexation capacity of raw and settled sewage via the solubilization of Cu^{2+} from the precipitate has been developed.[64] In fact, hydroxide and carbonate precipitates are readily solubilized by organic ligands, but sulfides, which are highly insoluble, may only be solubilized if oxidized to sulfates.[75] This suggests that if there is sufficient residual complexation capacity, precipitated forms of metals in sewage may be solubilized by the organic ligands present, whereas in anaerobic sludge digestion, a greater degree of insolubility will obtain, unless reoxidation of the predominant reduced species occurs.

The formation of significant quantities of what appeared to be heavy-metal precipitates in the particulate-free fraction of activated sludge has been demonstrated.[76] Samples of mixed liquor were filtered to remove suspended solids, soluble metal salts were added to the filtrate, and the samples were refiltered through 0.22-μm pore size membrane filters to remove the precipitates formed. At heavy-metal concentrations of up to 2×10^{-4} M, >88% of lead was precipitated, whereas trivalent chromium and copper precipitation was 53 to 95 and 49 to 87%, respectively, dependent on concentration. The concentrations added were approximately equivalent to or greater than the typical complexation capacities given in Table 1. Most of the cadmium, dichromate, cobalt, manganese, and nickel remained soluble.

The detailed study of precipitation is probably much more difficult than the study of soluble phase speciation, especially in a matrix such as activated sludge mixed liquor, which contains high concentrations of suspended solids. One approach for the study of particulates of all classes is membrane filtration using a range of pore sizes. Chen et al.[77] observed the preferential association of lead, manganese, and nickel with particles at the lower end of the range of sizes found in sewage effluent (<8 μm), whereas cadmium, chromium, and copper all tended to associate mainly with larger particles. It is possible that the smaller particles would represent in part the chemical precipitates, since insoluble lead phosphate has a mean particle diameter of 0.4 to 0.6 μm.[50] Kempton et al.,[54] in an attempt to resolve the particulate fractions in the mixed liquor and final effluent from an activated sludge simulation treating a soluble, synthetic sewage, found that the essential distinction between the solid and the soluble phase occurred for heavy metals associated with particle sizes of >20 μm and soluble species (defined as <0.2 μm): no fractionation of different particle classes in the intermediate range was observed. These data, which are in marked contrast to the results obtained using real sewage,[77] indicate that two fundamental physical forms of the heavy metals existed, either in association with the activated sludge flocs or in solution. In the activated sludge simulation, all of the suspended solids were of biological origin, which suggests that many of the discrete particle sizes observed in the full-scale study were components which had arisen from the original raw sewage.

This suggestion is further supported by work on a pilot-scale primary sedimentation pilot plant.[78,79] Discrete particle size fractions were found in the range 20 to 1000 μm. Copper, zinc, and lead tended to be concentrated in the fractions corresponding to particle diameters up to 35 μm and may have been present as inorganic precipitates, whereas silver, manganese, and chromium were associated to a greater degree with the larger organic particles.[79]

V. ENVIRONMENTAL SIGNIFICANCE OF SPECIATION

Florence[2] has stated that in the near future it is inevitable that water-quality legislation for heavy metals will include statements relating to their speciation and that it may be possible to tolerate higher total concentrations of some metals as long as the "labile" or bioavailable fraction is below a certain level. With respect to the disposal of sewage sludge to land, the importance of heavy-metal speciation has already been recognized insofar that it influences bioavailability.

Speciation is rarely performed in wastewater-treatment processes on an operational basis, although its importance has been recognized in anerobic wastewater treatment since this tends to be the biological process most susceptible to heavy-metal toxicity. Toxic effects may be ameliorated by adding sulfide and raising the pH and this would also have the effect of reducing metal mobility in the digested sludge produced.[80] It is feasible that similar control strategies could be applied to other unit processes.

Although a variety of methods have been used to characterize metal forms in aqueous samples, not all have been applied to wastewater matrices. Many methods would be clearly unsuitable due to the constraints imposed by interferences, selectivity, and sensitivity, whereas many others are difficult to validate due to the complex nature of the samples. Physical separation methods may afford the best opportunity for speciation wherein the effects of high suspended-solids concentrations and matrix interference may be overcome. A technique such as gel filtration, although considered unsuitable for natural waters, can be applied to the soluble phase of wastewater matrices by virtue of the higher concentrations of heavy metals present. However, physical separation techniques including filtration for studies of the important particulate phase, will only distinguish broadly defined groups of ligands and hence will not completely resolve the complexity of this type of sample. In such cases, the use of more refined interpretations of the analytical data to take account of the simultaneous formation of two or more complexes will be required. Since free metal ions appear to be present in sewage matrices at only very low levels, methods which determine only these species will be of limited use. In terms of the removal of heavy metals and their ultimate behavior in the receiving environment, the equilibrium between complexation by soluble and particulate ligands may be an important regulatory factor.

REFERENCES

1. **Buffle, J.**, Speciation of trace elements in natural waters, *Trends Anal. Chem.*, 1, 90, 1981.
2. **Florence, T. M.**, The speciation of trace metals in waters, *Talanta*, 29, 345, 1982.
3. **Hart, B. T.**, Trace metal complexing capacity of natural waters: a review, *Environ. Technol. Lett.*, 2, 95, 1981.
4. **Neubecker, T. A. and Allen, H. E.**, The measurement of complexation capacity and conditional stability constants for ligands in natural waters, *Water Res.*, 17, 1, 1983.
5. **Reuter, J. H. and Perdue, E. M.**, Importance of heavy metal organic matter interactions in natural waters, *Geochim.Cosmochim. Acta*, 41, 325, 1977.
6. **Forstner, U. and Wittmann, G. T. W.**, *Metal Pollution in the Aquatic Environment*, Springer-Verlag, New York, 1979.
7. **Anthony, R. M. and Breimhurst, L. H.**, Determining maximum influent concentrations of priority pollutants for treatment plants, *J. Water Pollut. Control Fed.*, 53, 1457, 1981.
8. **Lester, J. N.**, Significance and behaviour of heavy metals in waste water treatment processes. I. Sewage treatment and effluent discharge, *Sci. Total Environ.*, 30, 1, 1983.
9. **Lester, J. N., Sterritt, R. M., and Kirk, P. W. W.**, Significance and behaviour of heavy metals in waste water treatment processes. II. Sludge treatment and disposal, *Sci. Total Environ.*, 30, 45, 1983.
10. **Steiner, A. F., McLaren, D. A., and Forster, C. F.**, The nature of activated sludge flocs, *Water Res.*, 10, 25, 1976.
11. **Lawson, P. S., Sterritt, R. M., and Lester, J. N.**, Adsorption and complexation mechanisms of heavy metal uptake in activated sludge, *J. Chem. Technol. Biotechnol.*, 34B, 253, 1984.
12. **Rudd, T., Sterritt, R. M., and Lester, J. N.**, Complexation of heavy metals by extracellular polymers in the activated sludge process, *J. Water Pollut. Control Fed.*, 56, 1260, 1984.
13. **Rudd, T., Sterritt, R. M., and Lester, J. N.**, Stability constants and complexation capacities of complexes formed between heavy metals and extracellular polymers from activated sludge, *J. Chem. Technol. Biotechnol.*, 33A, 374, 1983.
14. **Ruzic, I.**, Theoretical aspects of the direct titration of natural waters and its information yield for trace metal speciation, *Anal. Chim. Acta*, 140, 99, 1982.
15. **Mantoura, R. F. C. and Riley, J. P.**, The use of gel filtration in the study of metal binding by humic acids and related compounds, *Anal. Chim. Acta*, 78, 193, 1975.
16. **Moody, T. D. and Thomas, J. D. R.**, *Selective Ion Sensitive Electrodes*, Merrow, Watford, U.K., 1971.
17. **Barica, J.**, Unusual response of a cupric ion electrode in prairie lake water, *J. Fish. Res. Board Canada*, 35, 141, 1978.
18. **El-Taras, M. F. and Pungor, E.**, The influence of some organic complexing agents on the potential of copper (II) selective electrodes. Application of the silicone rubber-based electrode to the determination of citrate and 8-hydroxyquinoline, *Anal. Chim. Acta*, 82, 285, 1976.
19. **Sillen, L. G. and Martell, A. E.**, *Stability Constants of Metal Ion Complexes*, Suppl. 1, Special Publ. No. 25, The Chemical Society, London, 1971.
20. **Sterritt, R. M. and Lester, J. N.**, Aspects of the determination of complexation parameters for metal-particulate complexes in activated sludge, *Water Res.*, 19, 315, 1985.
21. **Sterritt, R. M. and Lester, J. N.**, Influence of bacterial growth on the forms of cadmium in defined culture media, *Bull. Environ. Contam. Toxicol.*, 24, 196, 1980.
22. **Sterritt, R. M. and Lester, J. N.**, unpublished results, 1978.
23. **Astruc, M. and Mericam, P.**, personal communication, 1978.
24. **van de Meent, D., Los, A., de Leeuw, J. W., Schenck, I. A., and Salomons, W.**, Stability constants and binding capacities of fractionated suspended matter for cadmium, *Environ. Technol. Lett.*, 2, 569, 1981.
25. **Jardim, W. F. and Allen, J. E.**, Measurement of copper complexation by naturally occurring ligands, in *Complexation of Trace Metals in Natural Waters*, Kramer, C. J. M. and Duinker, J. C., Eds., Martinus Nijhoff/Dr. W. Junk, The Hague, 1984, 1.
26. **Gardiner, J.**, The chemistry of cadmium in natural water. I. A study of cadmium complex formation using the cadmium specific ion electrode, *Water Res.*, 8, 23, 1974.
27. **Cantwell, F. F., Nielsen, J. S., and Hrudey, S. E.**, Free nickel ion concentration in sewage by an ion exchange column-equilibration method, *Anal. Chem.*, 54, 1498, 1982.
28. **Cheng, M. H., Patterson, J. W., and Minear, R. A.**, Heavy metals uptake by activated sludge, *J. Water Pollut. Control Fed.*, 47, 362, 1975.
29. **Laxen, D. P. H. and Harrison, R. M.**, A scheme for the physicochemical speciation of trace metals in freshwater samples, *Sci. Total Environ.*, 19, 59, 1981.
30. **Crosser, M. L. and Allen, H. E.**, Complexation of heavy metals by ligands in industrial wastewater — measurement and effect on metals removal, *Proc. 32nd Industrial Waste Conf. Purdue Univ.*, Purdue University, West Lafayette, Ind., 345, 1977.

31. **Laxen, D. P. H. and Harrison, R. M.,** The physicochemical speciation of Cd, Pb, Cu, Fe, and Mn in the final effluent of a sewage treatment works and its impact on speciation in the receiving river, *Water Res.*, 15, 1053, 1981.
32. **Bender, M. E., Matson, W. R., and Jordan, R. A.,** On the significance of metal complexing agents in secondary sewage effluents, *Environ. Sci. Technol.*, 4, 520, 1970.
33. **Shuman, M. S. and Woodward, G. P.,** Stability constants of copper-organic chelates in aquatic samples, *Environ. Sci. Technol.*, 11, 809, 1977.
34. **O'Shea, T. A. and Mancy, K. H.,** Characterisation of trace metal organic interactions by anodic stripping voltammetry, *Anal. Chem.*, 48, 1603, 1976.
35. **Sposito, G.,** Trace metals in contaminated waters, *Environ. Sci. Technol.*, 15, 396, 1981.
36. **Tuschall, J. R. and Brezonik, P. L.,** Evaluation of the copper anodic stripping voltammetry complexometric titration for complexing capacities and conditional stability constants, *Anal. chem.*, 53, 1986, 1981.
37. **Sterritt, R. M. and Lester, J. N.,** Comparison of methods for the determination of conditional stability constants of heavy metal-fulvic acid complexes, *Water Res.*, 18, 1149, 1984.
38. **Castetbon, A., Corrales, M., Astruc, M., Dotin, M., Sterritt, R. M., and Lester, J. N.,** Comparative study of heavy metal complexation by fulvic acid, *Environ. Technol. Lett.*, 7, 495, 1986.
39. **Varney, M. S., Turner, D. R., Whitfield, M., and Mantoura, R. F. C.,** The use of electrochemical techniques to monitor complexation capacity titrations in natural waters, in *Complexation of Trace Metals in Natural Waters,* Kramer, C. J. M. and Duinker, J. C., Eds., Martinus Nijhoff/Dr. W. Junk, The Hague, 1984, 33.
40. **de Mora, S. J. and Harrison, R. M.,** The use of physical separation techniques in trace metal speciation studies, *Water Res.*, 17, 723, 1983.
41. **Mercer, E. R. and Richmond, J. L.,** An investigation of the chemical form of copper in soil solutions, *Agric. Res. Counc. (G. B.) Letcombe Lab. Annu. Rep.*, 19, 46, 1968.
42. **Mercer, E. R. and Richmond, J. L.,** An investigation of the chemical form of copper in soil solutions, *Agric. Res. Counc. (G. B.) Letcombe Lab. Annu. Rep.*, 20, 61, 1969.
43. **Steinberg, C.,** Species of dissolved metals derived from oligotrophic hard water, *Water Res.*, 14, 1239, 1980.
44. **Baham, J., Ball, N. B., and Sposito, G.,** Gel filtration studies of trace metal-fulvic acid solutions extracted from sewage sludges, *J. Environ. Qual.*, 7, 181, 1978.
45. **Sterritt, R. M. and Lester, J. N.,** Speciation of copper and manganese in effluents from the activated sludge process, *Environ. Pollut. (Ser. A).*, 27, 37, 1982.
46. **Lawson, P. S., Sterritt, R. M., and Lester, J. N.,** The speciation of metals in sewage and activated sludge effluent, *Water Air Soil Pollut.* 21, 387, 1984.
47. **Rossin, A. C., Sterritt, R. M., and Lester, J. N.,** The influence of process parameters on the removal of heavy meals in activated sludge, *Water, Air Soil Pollut.*, 17, 185, 1982.
48. **Florence, T. M. and Batley, G. E.,** Determination of the chemical forms of trace metals in natural waters, with special reference to copper, lead, cadmium and zinc, *Talanta*, 24, 151, 1977.
49. **Means, J. L., Crerar, D. A., and Amster, J. L.,** Application of gel filtration chromatography to evaluation of organometallic interactions in natural waters, *Limnol. Oceanogr.*, 22, 957, 1977.
50. **Posner, A. M.,** Importance of electrolyte in the determination of molecular weights by 'Sephadex' gel filtration with especial reference to humic acid, *Nature (London),* 198, 161, 1963.
51. **Brown, M. J.,** Metal Removal by Bacterial Extracellular Polymers in Activated Sludge, Ph.D. thesis, Imperial College of Science and Technology, University of London 1981.
52. **Stoveland, S. and Lester, J. N.,** A study of the factors which influence metal removal in the activated sludge process, *Sci. Total Environ.*, 16, 37, 1980.
53. **Hayes, T. D. and Theis, T. L.,** The distribution of heavy metals in anaerobic digestion, *J. Water Pollut. Control Fed.*, 50, 61, 1978.
54. **Kempton, S., Sterritt, R. M., and Lester, J. N.,** Factors affecting the fate and behaviour of toxic elements in the activated sludge process, *Environ. Pollut. (Ser. A),* 32, 51, 1983.
55. **Barnes, D., Cooksey, B. G., Metters, B., and Metters, C.,** Interferences in the electroanalytical determination of copper and lead in water and waste water, *J. Water Treat. Exam*, 24, 318, 1975.
56. **Benjamin, M. M. and Leckie, J. O.,** Conceptual model for metal-ligand-surface interactions during adsorption, *Environ. Sci. Technol.*, 15, 1050, 1981.
57. **Kirk, P. W. W., Lester J. N., and Perry, R.,** The behaviour of nitrilotriacetic acid during the anaerobic digestion of sewage sludge, *Water Res.*, 16, 973, 1982.
58. **Laxen, D. P. H. and Chandler, I. M.,** Comparison of filtration techniques for size distribution in fresh waters, *Anal. Chem.*, 54, 1350, 1982.
59. **Patterson, J. W. and Hao, S. S.,** Heavy metals interactions in the anaerobic digestion system, *Proc. 30th Industrial Waste Conf., Purdue University,* Purdue University, West Lafayette, Ind., 1979, 544.
60. **Gould, M. S. and Genetelli, E. J.,** Heavy metal complexation behaviour in anaerobically digested sludges, *Water Res.*, 12, 505, 1978.

61. **Cranston, R. E. and Buckley, D. E.**, The application and performance of microfilters in analysis of suspended particulate matter, *Bedford Institute Report*, BI-R-72-7, 14, 1972.
62. **Danielsson, L. G.**, On the use of filters for distinguishing between dissolved and particulate fractions in natural waters, *Water Res.*, 16, 179, 1982.
63. **Patterson, J. W.**, Parameters influencing metals removal in POTW's, *Proc. 8th Natl. Conf. Municipal Sludge Management*, Information Transfer, Silver Springs, Md., 1979, 82.
64. **Kunkel, R. and Manahan, S. E.**, Atomic absorption analysis of strong heavy metal chelating agents in water and wastewater, *Anal. Chem.*, 45, 1465, 1973.
65. **Hanck, K. W. and Dillard, J. W.**, Evaluation of micromolar compleximetric titrations for the determination of the complexing capacity of natural waters, *Anal. Chim. Acta*, 89, 329, 1977.
66. **Lewin, B. H. and Rowell, M. J.**, Trace metals in sewage effluent, *Effluent Water Treat. J.*, 13, 273, 1973.
67. **Brown, M. J. and Lester, J. N.**, Metal removal in activated sludge: the role of bacterial extracellular polymers, *Water Res.*, 13, 817, 1979.
68. **Brown, M. J. and Lester, J. N.**, Role of bacterial extracellular polymers in metal uptake in pure bacterial culture and activated sludge. I. Effects of metal concentration, *Water Res.*, 16, 1539, 1982.
69. **Brown, M. J. and Lester, J. N.**, Role of bacterial extracellular polymers in metal uptake in pure bacterial culture and activated sludge. II. Effects of mean cell retention time, *Water Res.*, 16, 1549, 1982.
70. **Rudd, T., Sterritt, R. M., and Lester, J. N.**, Formation and conditional stability constants of complexes formed between heavy metals and bacterial extracellular polymers, *Water Res.*, 18, 379, 1984.
71. **Forster, C. F.**, Factors involved in the settlement of activated sludge. II. The binding of polyvalent metals, *Water Res.*, 19, 1265, 1985.
72. **Sterritt, R. M. and Lester, J. N.**, The influence of sludge age on heavy metal removal in the activated sludge process, *Water Res.*, 15, 59, 1981.
73. **Saunders, F. M. and Dick, R. I.**, Effect of mean cell residence time on organic composition of activated sludge effluent, *J. Water Pollut. Control Fed.*, 53, 201, 1981.
74. **Boyd, S. A., Sommers, L. E., and Nelson, D. W.**, Infrared spectra of sewage sludge fractions: evidence for an amide metal binding site, *Soil Sci. Soc. Am. J.*, 43, 893, 1979.
75. **Huang, C. P. and Kao, J. F.**, The removal of heavy metals from municipal sludges, *Proc. 8th Natl. Conf. Municipal Sludge Management*, Information Transfer, Silver Springs, Md., 1979, 225.
76. **Sterritt, R. M., Brown, M. J., and Lester, J. N.**, Metal removal by adsorption and precipitation in the activated sludge process, *Environ. Pollut. (Ser. A)*, 24, 313, 1981.
77. **Chen, K. Y., Young, C. S., and Rohatgi, N.**, Trace metals in waste water effluents, *J. Water Pollut. Control Fed.*, 46, 2663, 1974.
78. **Kempton, S., Sterritt, R. M., and Lester, J. N.**, Heavy metal removal in primary sedimentation. I. The influence of metal solubility, *Sci. Total Environ.*, 63, 231, 1987.
79. **Kempton, S., Sterritt, R. M., and Lester, J. N.**, Heavy metal removal in primary sedimentation. II. The influence of metal speciation and particle size distribution, *Sci. Total Environ.*, 63, 246, 1987.
80. **Mosey, F. E.**, Assessment of the maximum concentration of heavy metals in crude sewage which will not inhibit the anaerobic digestion of sludge, *Water Pollut. Control*, 75, 10, 1976.

INDEX

A

Acetic acid, 130
Acid-base interactions, 8
Acid digestion, 107—111
Acid extractions
   discrete, 129—132
   progressive acidification techniques, 132—136
Acids, acceptor, 2—4
Activated sludge, 167—169, see also Analysis; Speciation
Adsorption, 5, 8, see also Speciation
   gel filtration chromatography, 163
   membrane filtration, 164—165
Adsorption-desorption equilibria, 133
Adsorption isotherms, 157—158
Aeration, and solubility, 128
Agricultural land, regulatory considerations, 67
Alkali earth metals, 8
Aluminum
   analysis, 108, 109
      precision data, 120
      technical considerations, 113—115, 117
   regulation
      U.S. standards, 92
      WHO and European standards, 69, 72
   sources
      atmospheric deposition, 39
      domestic discharges, 45, 46
      industrial, 45, 47, 50—53
      lithosphere, 34
      mining production, 36, 37
      by product type, 46
   speciation
      chemical extractions, 133
      distribution, 147
   toxicity and health effects, 19
Aluminum chlorohydrate, 115
Aluminum fluoride, 90
Amino acids, 146
Ammonia, complex formation, 156
Ammonium acetate, 129, 130, 139, 141
Anaerobic digestion, 8
Anaerobic wastewater treatment, speciation effects, 170
Analysis, see also Speciation
   atomic absorption spectrophotometry, 107—119
      electrothermal atomization, 112—116
      flame atomization, 107—112
      hydride generation, 118
      mercury, 116—118
      precision data, 118—119
   other methods, 119—121
   sampling, 106—107
Anode stripping voltammetry, 161—162, 166
Anthropogenic sources, 32—34
Antimony (Sb)
   analysis
      hydride generation, 118
      precision data, 120
      technical considerations, 113, 114
   regulation
      land disposal, 96
      U.S. standards, 87
      U.S.S.R. standards, 94
      WHO and European standards, 72, 73
   sources, 33
      atmospheric deposition, 38
      domestic, 46
      industrial, 47, 51
      lithosphere, 34
      mining production, 36, 37
Aqua regia, 110, 118
Aqueous extractions, 128
Arsenic
   analysis
      hydride generation, 118
      precision data, 120
      technical considerations, 113, 114
   biomethylation, 7
   properties, 1, 2
   regulation, 66
      comparison of guidelines, 100
      land disposal, 96
      sea disposal, 98
      U.K. standards, 76—77
      U.S. standards, 85, 87, 89, 92
      U.S.S.R. standards, 93, 94
      WHO and European standards, 69, 70, 72, 73, 75
   solubilities, 7
   sources
      atmospheric deposition, 38
      domestic, 46
      industrial, 47, 51, 52
   toxicity and health effects, 11—12
Ascorbic acid, 120
Ashing techniques, 111
Atmospheric deposition, 38—40, 42—43
Atomic absorption spectrophotometry, 136, see also Analysis

B

Bacteria, see also Biological processing
   distribution and, 147
   extracellular polymers, 167
Barium
   regulation
      U.S. standards, 85, 87, 92
      WHO and European standards, 69, 70, 72, 73
   solubilities, 7
Barium chloride, 129, 142
Beryllium
   analysis

ashing temperatures, 116
  precision data, 120
  technical considerations, 113, 114
 regulation, 66—67
  land disposal, 96
  U.S. standards, 87, 92
  U.S.S.R. standards, 93, 94
  WHO and European standards, 69, 73
 solubilities, 7—8
 sources
  atmospheric deposition, 39
  domestic, 46
  industrial, 47, 51
  lithosphere, 34
  mining production, 35—37
 toxicity and health effects, 19
Binding capacity, activated sludge, 167
Bioavailability
 oxidation state and, 4—6
 properties affecting, 7, 8
Biological processing
 anaerobically digested, 168
 extracellular polymers, 167
 regulatory considerations, 67
 speciation effects, 170
Biomethylation, see Methylation
Bismuth
 analysis
  ashing temperatures, 116
  hydride generation, 118
  precision data, 120
  technical considerations, 113, 114, 117
 sources
  domestic, 46
  industrial, 47, 51
  lithosphere, 34
  mining production, 36, 37
Black list, 71—73
Borohydride, 118
Boron regulation
 land disposal, 96
 U.S. standards, 92
 WHO and European standards, 70, 72, 73
Borosilicate glass, 107
Bromination, 117

C

Cadmium
 analysis, 107, 109—111
  ashing temperatures, 116
  electrothermal ashing, 112
  other methods, 121
  precision data, 120
  sample injection, 115
  technical considerations, 113, 114, 117
 biomethylation, 7
 chemical properties and speciation, 5, 6
 properties and speciation, 8
 regulation, 66
  comparison of guidelines, 100
  EEC directives, 74
  land disposal, 95, 96, 98
  sea disposal, 98
  U.K. standards, 76
  U.S. standards, 85, 87, 89, 91, 92
  U.S.S.R. standards, 94
  WHO and European standards, 69, 70, 72, 75
 sources, 33, 53—58
  atmospheric deposition, 38—40
  domestic wastewater, 46
  industrial uses, 47, 51
  lithosphere, 34
  mining production, 35—37
  runoff, 43, 44
 speciation, 156
  chemical extractions, discrete, 128, 129, 131—135
  chemical extractions, sequential, 138, 141, 142
  distribution, 143—147
  gel filtration chromatography, 162—163
  ion-specific electrodes, 159
  occurrence in wastewaters, 166—170
  toxicity and health effects, 9—10, 12—13
  zinc-cadmium pairs, 129
Cadmium-zinc ratios, 18
Calcium
 analysis, 109
  precision data, 120
  technical considerations, 113—117
 biological systems, heavy metal replacement of, 11
 complex formation, 156
 properties and speciation, 8
Calcium chloride, 129
Carbonates, 138—142, 147
 complex formation, 156
 distribution, 143, 145—147
 EDTA and, 142
Carcinogenic risk, 6, 66, 73, see also Toxicity, human
Carry-over, 136
Case studies, 55—58
Cation-exchange processes
 chemical extractions, 133
 zinc binding, 129
Cation exchange resins, 159—160
Cationic polyelectrolytes, 115—116
Chelates, see Speciation
Chelating agents, see also DTPA; EDTA
 analytic procedures, 107
 extractions with, 129—132
Chelex, 166
Chemical extractions, see Speciation, chemical
Chemical oxygen demand, 163
Chemical properties, 4—8, 126—127
Chemical speciation, see Speciation, chemical
Chlor-alkali, 50, 73—74, 90
Chloride extractions, 129
Chlorides
 complex formation, 156

gel filtration chromatography, 163
Chromatography
　cation exchange, 159—160
　gel filtration, 162—164, 169
Chromium
　analysis, 107, 109—111
　　electrothermal ashing, 112
　　precision data, 120
　　sample injection, 115, 116
　　technical considerations, 113, 114, 117
　regulation, 66
　　comparison of guidelines, 100, 101
　　land disposal, 95, 96, 98
　　sea disposal, 98, 99
　　U.K. standards, 77—78
　　U.S. standards, 85, 87, 89, 92
　　U.S.S.R. standards, 94
　　WHO and European standards, 69, 70, 72, 73, 75
　sources, 33, 53—58
　　atmospheric deposition, 39
　　domestic, 46
　　fossil fuels, 50
　　industrial, 50—53
　　lithosphere, 34
　　mining production, 36, 37
　　runoff, 43, 44
　speciation, 156
　　chemical extractions, 128, 131—134
　　distribution, 147
　　occurrence in wastewaters, 169, 170
　toxicity and health effects, 16—17
Chromium pigments, 90
Citrate, 159
Citric acid, 130, 131
Classifications, 2—4
Clay minerals, 126
Cleaning compounds, 43
Clostridium cochlearium, 7
Cobalt
　analysis, 108
　　precision data, 120
　　technical considerations, 113, 114, 117
　regulation
　　land disposal, 96
　　U.S. standards, 92
　　U.S.S.R. standards, 94
　　WHO and European standards, 69, 73
　sources
　　atmospheric deposition, 39
　　domestic, 46
　　industrial, 48, 50, 51, 53
　　lithosphere, 34
　　mining production, 35—37
　speciation, 156
　　chemical extractions, 128, 131—134
　　occurrence in wastewaters, 167—169
Colloids, 5
Complexes, 5
Complex formation, 5—6, 8, see also Analysis; Speciation

　theoretical basis, 157—158
　types of, 137, 139—140
Copper
　analysis, 107—111
　　electrothermal ashing, 112
　　precision data, 120
　　sample injection, 115, 116
　　technical considerations, 113, 114, 117
　chemical properties and speciation, 5—6
　properties and speciation, 8
　regulation
　　comparison of guidelines, 100, 101
　　land disposal, 95, 96, 98
　　sea disposal, 99
　　U.K. standards, 79—80
　　U.S. standards, 85, 87, 89, 90, 92
　　U.S.S.R. standards, 93, 94
　　WHO and European standards, 69, 70, 72, 73, 75
　sources, 33, 53—58
　　atmospheric deposition, 38, 40
　　domestic, 44—46
　　industrial, 48, 51, 52
　　lithosphere, 34
　　mining production, 36, 37
　　runoff, 40, 41, 43, 44
　speciation, 156
　　anodic stripping voltammetry, 162
　　chemical extractions, discrete, 129, 131—135
　　chemical extractions, sequential, 138, 139, 141, 142
　　distribution, 144—147
　　gel filtration chromatography, 163, 164
　　ion-specific electrodes, 159
　　occurrence in wastewaters, 166—170
　toxicity and health effects, 18—19
Copper hydroxide, 169
Copper-lead pairs, 129
Copper sulfate, 90
Coprecipitation, 8
Covalent bonding, 1, 2
Covalent ligands, 8
Cysteine, 146

D

DC argon plasma atomic emission spectrophotometry, 136
Decon 90, 107
Deicing compounds, 43
Dichromate, 163
Diethylenetriaminepentaacetic acid, 128
Differential pulse anodic stripping voltammetry, 120, 162
Dipole-dipole interactions, 126
Domestic discharges, 44—46
Dowex 50W-X-8, 159
DPTA, 130—131, 139, 141
Drinking water standards, see Legislation
Dry ashing, 111

E

EDTA, 128
  analysis, acid leaching, 107
  carbonates and, 142
  chemical extractions, discrete, 130—131, 137
  chemical extractions, sequential, 140—142
EDTA-extractable species, see Carbonates
Electrolysis, see Chlor-alkali
Electron acceptors, 1, 2
Electronegativity, and bioavailability, 8
Electroplating, 50, see also Chlor-alkali
Electrostatic charge, and adsorption, 126
Electrothermal atomization, 112—116
Emission spectrophotometry, 119
Environmental considerations, see Legislation
Enzyme systems, toxicity mechanisms, 10—11
European Economic Community directives on water quality
  cadmium discharges, 74
  chlor-alkali industry, 73—74
  drinking water, 70—71
  groundwater protection, 74—75
  for human consumption, 71
  pollution, 71—73
  U.K. environmental quality standards, 75—84
Exchangeable forms, distribution, 143—147
Exchangeable phase, 137—141
Exhaust emission, 42—43
Extracellular polymers, bacterial, 167
Extractions, see Speciation

F

Federal regulation, see Legislation
Ferrous sulphate, 115
Filtration, membrane, 164—166
Flame atomization, 107—112
Fluorescence spectroscopy, X-ray, 121, 136
Fossil fuels, heavy metals in, 50
Freundlich isotherm, 157
Fuels, 42—43, 50
Fulvic acid, 8, 126, 162, 163

G

Gel filtration chromatography, 162—164, 169
Glycine, 159
Gold mining, 50
Government standards, see Legislation
Gray list, 71—73
Groundwater, regulatory considerations, 67

H

Hard acids, 1, 2
Health effects, see Toxicity, human
Humic acid, 8, 126
Humic material, 142
Hydrated ions, 5
Hydride generation, 118
Hydrochloric acid, 110
Hydrofluoric acid, 90, 107—109
Hydrogen cyanide, 90
Hydrogen peroxide, 107, 110—112, 138—141
Hydrous oxides, 133
Hydroxides, 132, 142, 147
Hydroxylamine sulfate, 117

I

Inductively coupled plasma spectrophotometry, 121
Industrial discharges, 45—58
  case studies, 53—58
  metal uses, 45, 47—49
  relative contributions, 53
  selected industries, 50—53
Infrared spectroscopy, 147
Inks, 52—53
Inorganic chemicals manufacture, 50
Inorganic complexes, 5, 8
  analysis, see Analysis; Speciation
  solubilities, 6—8
Ion-dipole forces, 126
Ion-exchange processes
  adsorption by, 126
  chemical extractions, sequential, 137
Ion-exchange techniques, 159—161
Ionic radii, 8
Ionic strength, 8, 126
Ion pair extractions, 129
Ions, 5
Ion-specific electrodes, 119, 158—159
Iron
  analysis, 109, 120
    precision data, 120
    technical considerations, 113, 114, 117
  complex formation, 156
  properties and speciation, 8
  regulation
    U.S. standards, 85, 87, 92
    U.S.S.R. standards, 93, 94
    WHO and European standards, 69, 70, 72
  solubility of salts, 7
  sources
    domestic, 45, 46
    industrial, 51, 52
    lithosphere, 34
    mining production, 36, 37
    runoff, 43
  speciation
    chemical extractions, 133
    distribution, 147
Irrigation water, standards for, 99—101
Isotherms, 157—158
Isotope dilution analysis, 121
Itai-itai disease, 9—10

L

Labile complexes, 161, 162, 170
Land use, regulatory considerations, 67
Langmuir isotherm, 157
Lead
  analysis, 107, 109—111
    ashing temperatures, 116
    electrothermal ashing, 112
    other methods, 121
    precision data, 120
    technical considerations, 113, 114, 117
  biomethylation, 7
  chemical properties and speciation, 6
  copper-lead pairs, 129
  regulation, 66, 101
    comparison of guidelines, 100
    land disposal, 95, 96, 98
    sea disposal, 99
    U.K. standards, 80—81
    U.S. standards, 85, 87, 89, 90, 92
    U.S.S.R. standards, 93, 94
    WHO and European standards, 69, 70, 72, 73, 75
  sources, 33, 53—58
    atmospheric deposition, 38—40
    domestic, 44—46
    exhaust emissions, 42—43
    geochemical, 38
    industrial, 48, 50—53
    lithosphere, 34
    mining production, 36, 37
    runoff, 40, 41, 44
  speciation, 6, 156
    chemical extractions, discrete, 128, 129, 131—135
    chemical extractions, sequential, 138, 141, 142
    distribution, 143—145, 147
    ion-specific electrodes, 159
    occurrence in wastewaters, 166—168, 170
  toxicity and health effects, 13—14
Legislation, see also European Economic Community directives
  sludge disposal
    guidelines vs. water quality for irrigation, 99—101
    to land, 94—98
    to sea, 98—99
  water quality criteria and standards, 67—98
    European Economic Community directives, 67—84
    U.S. Government, 84—90
    U.S.S.R. Ministry of Health, 90—94
    World Health Organization, 67—68
Lewis acids, 1, 2
Ligands, 5
  analysis, see Analysis; Speciation
  binding of, 157—158
    protonation of, 165
  surface-bound vs. soluble, 159

Lithium, 92
Lithosphere, 34
Lubricants, 42

M

Macromolecular ligands, binding by, 158
Magnesium
  analysis, 108, 109
    precision data, 120
    technical considerations, 113—115, 117
  biological systems, heavy metal replacement of, 11
  complex formation, 156
  properties and speciation, 8
  speciation, 156
Magnesium chloride, 129
Manganese
  analysis, 110
    precision data, 120
    technical considerations, 113, 114, 117
  regulation
    U.S. standards, 85, 87, 92
    U.S.S.R. standards, 93, 94
    WHO and European standards, 69, 70, 72
  sources
    atmospheric deposition, 38, 39
    domestic wastewater, 46
    industrial, 48, 50, 51
    lithosphere, 34
    mining production, 36, 37
    runoff, 43
  speciation, 156
    chemical extractions, discrete, 131—132
    chemical extractions, sequential, 138, 141, 142
    distribution, 147
    gel filtration chromatography, 163
    occurrence in wastewaters, 168—170
Marine environment, regulatory considerations, 67
Mass spectrometry, 121
Matrices, see Analysis
Membrane filtration, 164—166
Mercury
  analysis, atomic absorption spectrophotometry, 116—118
  biomethylation, 7
  properties and speciation, 8
  regulation, 66
    EEC directives, 73—74
    land disposal, 96
    sea disposal, 98
    U.S. standards, 85, 87, 89, 90, 92
    U.S.S.R. standards, 94
    WHO and European standards, 69, 70, 72
  solubility of salts, 7
  sources, 33, 56—58
    atmospheric deposition, 38, 39
    domestic, 46
    industrial, 49—51
    lithosphere, 34

mining production, 36, 37
speciation, 156
toxicity and health effects, 9—10, 14—15
Metal finishing, 50, 52
Metal ion pairs, 129
Metalloids, 1
Methionine, 146
Methylation, 7, 8, 169
Microbial action, and distribution, 147
Microorganisms, extracellular polymers, 167
Minimata disease, 10
Mining, 34—37
Molybdenum
  analysis, 116
    precision data, 120
    technical considerations, 113, 114, 117
  regulation
    land disposal, 96, 98
    U.S. standards, 92
    U.S.S.R. standards, 93
    WHO and European standards, 69, 73
  sources
    domestic, 46
    industrial, 49, 51, 53
    lithosphere, 34
    mining production, 36, 37
Mytilus edulis, 74

N

Neutron activation analysis, 121
Nickel
  analysis, 107, 109, 110
    electrothermal ashing, 112
    other methods, 121
    precision data, 120
    sample injection, 115
    technical considerations, 113, 114, 117
  chemical properties and speciation, 4—5
  regulation
    comparison of guidelines, 100, 101
    land disposal, 96, 98
    sea disposal, 99
    U.K. standards, 82, 95
    U.S. standards, 87, 90, 92
    U.S.S.R. standards, 94
    WHO and European standards, 72, 73, 75
  sources, 53—58
    atmospheric deposition, 39
    domestic, 44, 46
    industrial, 49—52
    lithosphere, 34
    mining production, 36, 37
    runoff, 43, 44
  speciation, 4—5, 156
    chemical extractions, discrete, 128, 131—136
    chemical extractions, sequential, 138, 141, 142
    distribution, 144—147
    gel filtration chromatography, 163
    ion exchange methods, 159

occurrence in wastewaters, 166—170
toxicity and health effects, 15—16
Nickel sulfate, 90
Nitrates, gel filtration chromatography, 163
Nitric acid
  analytical procedures, 108—112, 115, 117
  chemical extractions, sequential, 138, 140—143
Nitric-acid extractable metals, see Sulfides

O

Organic complexes, 5, 8, 142, see also Analysis; Speciation
  distribution, 145, 146
  dry ashing, 111
Oxalate formation, 139
Oxidation-reduction potential, 127, 132
Oxidation states, 5
Oxides, 132, 138, 140, 141, 147, see also Speciation
  hydrous, 133
Oxidizable phase, 138, 139
Oxygen plasma ashing technique, 111

P

Paints, 52—53
Perchloric acid, 107, 112
Perchorate ion, 162
Permanganate, 117
Peroxides, 107, 110—112, 138—141
Persulfate, 117
pH, 8
  adsorption and, 126
  analysis, 107
  chemical extractions, chloride, 129
  distribution and, 147
  ligand protonation and, 165
  membrane filtration and, 165
  progressive acidification techniques, 132—136
  runoff and, 43
  solubility and, 128
  speciation and, 127
  sulfide solubility and, 137
  toxicity and, 170
Phosphates, 142, 147, 156
Phosphoric acid, 75
Photon activation analysis, 121
Photoprocessing, 52
Physical properties, and speciation, 126—127
Phytotoxic effects, 136
Pigments, 47—49, 52—53, 75, 90, 91
Plants, 86, 136
Plasma spectrophotometry, inductively coupled, 121
Polarography, 119—120, 161—162
Pollution, see Legislation
Polyelectrolytes, cationic, 115—116
Polyethylene, 107
Polymers, 5

bacterial, 167
Polypropylene, 107
Potassium fluoride, 137
Potassium permanganate, 117
Potassium persulfate, 117
Precipitates, 8, 132, 169—170, see also Speciation
Progressive acidification techniques, 132—136
Properties, 4—8
Protonation, ligand, 165
pT ranges, 9
Pyrophosphates, 137, 142

R

Rate constants, 161
Recovery, 117—118
Redox potential, 127, 132
Reducible phase, 138—141
Regulation, see Legislation
Relative standard deviation, 118—120
Residual phase, 140, 141
Road runoff, 41—43
Rock salt, 43
Runoff, 40—44

S

Sandy loams, 134, 135
Scatchard equation, 158
Selenium
  analysis, 112, 120
    hydride generation, 118
    precision data, 120
  biomethylation, 7
  regulation, 66
    land disposal, 96
    U.S. standards, 85, 87, 91, 92
    U.S.S.R. standards, 93, 94
    WHO and European standards, 69, 70, 72, 73
  sources
    domestic wastewater, 46
    geochemical, 38
    industrial, 49, 51
    lithosphere, 34
    mining production, 36, 37
  toxicity and health effects, 17—18
Sephadex resins, 163, 164, 169
Serine, 159
Silver
  analysis
    ashing temperatures, 116
    electrothermal ashing, 112
    precision data, 120
    technical considerations, 113, 114, 117
  regulation
    U.S. standards, 85, 87, 92
    U.S.S.R. standards, 93
    WHO and European standards, 69, 72, 73
  sources
    atmospheric deposition, 38
    domestic, 46
    industrial, 49, 51, 52
    lithosphere, 34
    mining production, 36, 37
  speciation, occurrence in wastewaters, 170
  toxicity and health effects, 20
Sodium bisulfate, 90
Sodium borohydride, 118
Sodium dichromate, 90
Soft acids, 1, 2
Soil type, 134, 135, 147
Solubility, 6—8, see also Speciation, chemical
  aeration and, 128
  distribution and, 143—147
  sulfides, 137
Sources
  anthropogenic and nonanthropogenic, 33—37
  atmospheric deposition, 38—40
  case studies, 55—58
  domestic discharges, 44—46
  geochemical background, 37—38
  industrial discharges, 45, 47—53
    metal uses, 45, 47—49
    types of industries, 50—53
  lithosphere, occurrence in, 34
  mine production, 34—37
  relative contributions of different sources, 53—54
  runoff, 40—44
  total environmental inputs, 33—34
Speciation, chemical, 4—8, see also Analysis
  discrete chemical extractions, 127—136
    acid reagents and chelating agents, 129—132
    aqueous, 128
    chloride, 129
    progressive acidification techniques, 132—136
  sequential chemical extractions, 136—147
    distributions of metals, 143—147
    problems of, 140—143
    techniques and reagents, 136—140
Speciation, physical and electrochemical
  analytical approach, 158—166
    gel filtration chromatography, 162—164
    ion exchange, 159—161
    ion-specific electrodes, 158—159
    membrane filtration, 164—166
    physical separations, 162—166
    polarography, 161—162
    techniques, 158—162
  environmental significance, 170
  species in wastewaters, 166—170
  theory, 157—158
Spectrophotometry, 121, see also Analysis
Spectroscopy, X-ray fluorescence, 121, 136
Stability constants, 159, 160, 165
Standard deviation, 118—120
Standards, see Legislation
Storm events, 43—44
Strontium, 93, 94
Sulfides
  solubility, 137

speciation
    chemical extractions, sequential, 138—141
    distribution, 144—147
    metal species distributions, 143
    precipitation, 8
    toxicity and, 170
Sulfur, toxicity effects, 10—11
Sulfuric acid, 107—109, 111, 117
Sulfur-oxidizing bacteria, 147
Sulfur sequestration, 1
Surface-bound ligands, 159

T

Tartrate, 159
Tellurium
    analysis
        hydride generation, 118
        precision data, 120
        technical considerations, 113, 114
    biomethylation, 7
    regulation
        U.S.S.R. standards, 94
        WHO and European standards, 73
    sources
        industrial, 49, 51
        lithosphere, 34
        mining production, 35, 36
    toxicity and health effects, 20
Temperature, and speciation, 127
Tetraethyllead, 42—43
Thallium
    analysis, 111
        other methods, 121
        precision data, 120
        technical considerations, 114, 117
    regulation
        U.S. standards, 87
        WHO and European standards, 73
    solubility, 8
    sources, 33
        industrial, 49, 51
        lithosphere, 34
        occurrence in wastewaters, 169
    toxicity and health effects, 19—20
Tin
    analysis
        precision data, 120
        technical considerations, 113, 114, 117
    biomethylation, 7
    regulation
        land disposal, 96
        WHO and European standards, 69, 73
    sources, 33
        domestic, 45, 46
        industrial, 49—52
        lithosphere, 34
        mining production, 35—37
Tires, 42
Titanium
    regulation
        U.S.S.R. standards, 94
        WHO and European standards, 73
    sources, domestic wastewater, 46
Titanium dioxide, 90
Total organic carbon, 156, 163
Toxicity, human
    aluminum, 19
    arsenic, 11—12
    beryllium, 19
    cadmium, 12—13
    chromium, 16—17
    copper, 18—19
    lead, 13—14
    mercury, 14—15
    nickel, 15—16
    regulatory considerations, 67
    selenium, 17—18
    silver, 20
    tellurium, 20
    thallium, 19—20
    vanadium, 20—21
    zinc, 18
Toxicity, microbial, 170
Toxicity, plants, 136
Tungsten, 94

U

Uranium, 69, 73
United States government water quality criteria, 84—90
U.S.S.R. water quality criteria, 90—94

V

Valence states, see Speciation
Vanadium
    analysis
        ashing temperatures, 116
        precision data, 120
        technical considerations, 113, 114, 117
    regulation
        land disposal, 96
        U.S. standards, 92
        U.S.S.R. standards, 94
        WHO and European standards, 69, 73
    sources, 49
        atmospheric deposition, 38, 39
        domestic, 46
        fossil fuels, 50
        lithosphere, 34
        mining production, 36, 37
    toxicity and health effects, 20—21
Van der Waals forces, 126
Volatility, methylated compounds, 7
Voltammetry, 119—120, 161—162, 166

W

Wastewater, see Analysis; Sources; Speciation
Water quality standards, see Legislation
World Health Organization water quality standards, 67—68

X

X-ray fluorescence spectroscopy, 121, 136

Z

Zinc
   analysis, 107—110
      precision data, 120
      technical considerations, 113, 114, 117
   chemical properties and speciation, 5, 6
   properties and speciation, 8
   regulation, 101
      comparison of guidelines, 100
      land disposal, 95, 96, 98
      sea disposal, 99
      U.K. standards, 83—84
      U.S. standards, 85, 87, 91, 92
      U.S.S.R. standards, 93, 94
      WHO and European standards, 69, 70, 72, 73, 75
   sources, 33, 49, 53—58
      atmospheric deposition, 38—40
      domestic, 44—46
      fossil fuels, 50
      geochemical, 37—38
      industrial, 50, 52, 53
      lithosphere, 34
      mining production, 36, 37
      runoff, 40—44
   speciation, 8, 156
      chemical extractions, discrete, 128, 129, 131—136
      chemical extractions, sequential, 138, 141, 142
      distribution, 144—147
      occurrence in wastewaters, 170
   toxicity and health effects, 18
Zinc-cadmium pairs, 129